OPTIMAL ILLUSIONS

OPTIMAL
ILLUSIONS

The False Promise
of Optimization

COCO KRUMME

RIVERHEAD BOOKS
NEW YORK 2023

RIVERHEAD BOOKS
An imprint of Penguin Random House LLC
penguinrandomhouse.com

ISBN 9780593331118 (hardcover)
ISBN 9780593331132 (ebook)

Printed in the United States of America
1st Printing

Book design by Daniel Lagin

CONTENTS

OPTIMAL ILLUSIONS

INTRODUCTION

IN THE MIDDLE OF THE FIELD, THE BULLDOZER PAUSES, AND OUT pops its driver for a cigarette break. It's a perfect September day in northern Kentucky: crisp, blue, free. A neighboring dozer carves ant-hill ruts through the construction site. In the distance are horse pastures and strip malls.

I weave past the open gate up one of the bulldozed roads until I reach the end of an ant rut. I park my small pickup truck, windows down, and let the dog out for a walk. The sun warms my face and my thoughts are lost in the dirt canvas, until suddenly, I feel the driver's eyes fix on me. This isn't my anthill to explore. He crushes out a cigarette and settles his gaze again in my direction, indifferent now, before turning back toward his rig and the work ahead.

Some say America's too new a country to have ruins. I say, our ruins are just better hidden, under the sheen of the new.

In Athens, the Parthenon stands tall and crumbling, like a

drive-in movie screen strung high above the tourists and café dwellers below. In Rome, the Colosseum anchors a swirling city, its cold stone presence cutting through the chatter of tour guides and the grease of cheap pizza and the plastic trinkets made overseas. At Teotihuacán, you can clamber up each imposing pyramid and practically touch the scorching sky.

In America, by contrast, we hide our ruins in the desert or cart them off to Chinese scrapyards. After September 11, the remnants of the World Trade Center were wrapped like widows in black canvas, quietly disassembled, and sent abroad. An engineer named Cao Xianggen says to the *Chicago Tribune*, of the injured steel: "America can't use it all, but China has a huge demand."

Now I watch the bulldozer driver carve new ruins into his field. He's helping build the Amazon Air Hub, a $1.5-billion site expected to host some one hundred cargo airplanes, three hundred trucks, and a robotic sort center sprawling over a million square feet. Slated to open in 2021, it's currently a mess of construction equipment and barricades.

As Americans moved more of their shopping online at the onset of the 2020 pandemic, Amazon sales increased some 40 percent. The company's revenues grew by a mind-boggling $100 billion. An efficiently scheduled fleet of aircraft will cut hours and days off already fast delivery times.

If the North America circulatory system conveys packages and goods to its every extremity, this corner of Kentucky represents its anthill heart. UPS and DHL and Wayfair have distribution centers here. Sixty-five percent of the population of the United States lives within six hundred miles: a one-day drive or ninety-minute flight away. It's close to five other airports, to rail lines and rivers and ports. Nearly three billion pounds of cargo pass through the Cincinnati/ Northern Kentucky International Airport annually, a number that continues to rise as Amazon expands its fleet. Something as seemingly local as a Maine lobster might be diverted west through Kentucky, only to be sent east again to an upscale restaurant in New York.

The circulatory system, its trucks and airplanes and conveyor belts and plastic bubble wrap, is kept in motion by a set of equations and code. This code tells each box where to go and when, in the most efficient way. Undergirding all this is a set of mathematical models that's explored each possibility, assigned probabilities, and planned contingencies. And beneath this set of models, the hidden choreographer, is a way of looking at the world.

THIS WAY OF SEEING IS THE SUBJECT OF THIS BOOK. IT'S A PERSPECtive that assumes our material reality can be understood, parceled, harnessed, and ultimately predicted. Not just Amazon's cargo jets but everything from the crashing of waves to the outbreak of war to the intricacies of love boils down to a set of parameters that can, with the right equipment, be measured and described.

It's an intoxicating thought and one that tugged at me for years. I was drawn to mathematical modeling because it promised an elegant bridge between pure logic and the messiness of the real world. Like the bulldozer carving lines on its anthill canvas, it offered the driver's seat, the chance to superimpose topography onto the rugged landscape. It also afforded a vantage point from which to marvel at what lay beyond, the world yet unmapped and perhaps not fully mappable.

The man I saw in the bulldozer was, in all likelihood, not fitting his ruts to the existing undulations of the construction site but carrying out a design engineered and approved in some faraway office, to make the most profitable use of the space at hand. Optimization enabled this, as well: shoving dirt according to a master plan, forcing the landscape into preconceived lines.

For a moment, I found myself quite unexpectedly at the pinnacle of this way of seeing. By the time I emerged from a stint at MIT and headed toward Silicon Valley, these places had become overnight meccas for optimization. Modern scientific computing was in full swing. The machinery for moving dirt was developing at breakneck speed. A new generation of cloud technologies spewed forth trillions of calculations per second, and buzzwords like "deep learning" and "neural networks" hung thick in the air like artificial fog at a rock concert. Software engineers shoved data into models, increasing efficiency in everything from digital advertisements to nutritional supplements.

Caught off guard at this pinnacle, but with the right skills, I decided to do two things at once. First, I dove into the foggy absurdity, working at start-ups and then striking out on my own. Second, I tried to maintain my romanticism about mathematics as a map, an elegant approximation. I failed at first because, unlike most people who succeed in Silicon Valley, I didn't care enough for money or status to chase it with any gusto. As for the second, well, no one seemed to care much for maps and topography when the money was falling off trees.

My own romanticism persisted. I started a scientific consulting business, working with researchers in logistics, climate science, and agriculture. I built models for how species evolve through a stretch of history, for how people's habits change, for the way a storm develops and how crops grow, for the movement of ships around the globe. I made maps and shoved dirt. Everyone wanted more data, more models, more solutions. The late June fog over the San Francisco Bay thickened. It was unsettling but also, in an altogether different way, intoxicating.

Except for one thing. I didn't buy it anymore.

My romanticism had waned. I increasingly felt the fakeness in all this abstraction, in trying to represent the lushness of a field with green pixels on a satellite map. By contrast, the general excitement for the tools of optimization seemed only to grow. The more I felt this fanaticism, the more I disbelieved it myself. My disenchantment was grounded in the excesses of the tech industry, but it didn't begin or end there. I bemoaned the yoke of corporate schedules and resisted upgrading my old flip phone for about a decade past the point of embarrassment. I began to contemplate, on the one hand, how to escape Silicon Valley and, on the other, how to destroy it.

Disenchanted and without any better idea, I started talking to people. I found myself wandering around, first in the agricultural Midwest and then all over the country. I met farmers and freight brokers, carpenters and oil workers. As soon as I could afford it, I bought a patch of land on a remote northwestern island with a run-down cabin and endeavored to learn the skills necessary to fix it up.

Everyone I told in San Francisco thought I was nuts. We were at the height of an incredible age, they reminded me. Not only that, but what I did was in high demand. Data-science salaries were through the roof. I had the right knowledge to make a mark. What was I doing rambling on about well-water testing and passive solar? "I don't understand, with your skills, why you're not flying around with billionaires in private jets right now," a cousin volunteered, only half joking. "The only way I can explain it"—it being my presence in San Francisco with a boatload of seemingly critical expertise and little interest in capitalizing—"is that you must be an ethical person or something," a venture capitalist said to me, as if he'd just encountered a polar bear in the Amazonian rain forest.

Indeed, as my peers were spinning up data teams at hedge funds and pharmaceutical companies, I was hanging out at sugar-beet factories in North Dakota. I still couldn't pinpoint the locus of my disillusionment, but I knew it had to do with the growing distance between my two worlds: the world represented by numbers on computer screens and the world in plain sight.

It wasn't until I moved into the half-built cabin, with a hole in the subfloor and walls I needed to finish and logs I needed to split in order to keep the wood stove going, that it really hit home. Here I was, fiddling away on my laptop for some far-off research group that wanted a better way to model wildfire spread in California so it could sell more insurance to homeowners. Then I'd put the laptop away and fiddle with the faulty well pump or stuff walls with the sheep's-wool insulation I'd carted up in cardboard boxes from Oregon. My vegetable starts mostly faltered, and I slowly made friends in the community. I realized: the skills that had brought me here, the skills that put me at the pinnacle of this new belief system of efficiency and optimization . . . these weren't the skills that would help me now.

THIS DISILLUSIONMENT WITH EFFICIENCY'S OVERREACH IS, I'LL argue, shared by many today, even those who don't dabble in optimization as I do. It surfaces often as a feeling of malaise: when the overwhelm of daily life and filled-to-the-brim personal calendars butts up against those questions that optimization can't solve. Am I doing the right thing? How do I even know what the right thing is? It manifests in sky-high rates of depression and anxiety, in the growing cognizance of breakdowns in supply chains and society, in migrations out of expensive cities and corporate jobs and even marriages.

Optimization is a way of seeing the world that has overtaken most others, in both modern America and much of the West. In fact, it has the peculiarity (an advantageous peculiarity, in terms of its own ability to persist) of rapidly crowding out others. Over the past century, optimization has made an impressive epistemic land grab. Take, for example, the dispiriting overreach of the MBA's vocabulary, of trade-offs and time management and incentives. Does your forecasting model fall short? Is it failing to account for externalities? No problem, we'll just update the top-right quadrant to incorporate what's left out. Every critique of the model is redirected toward improving the model itself. No one questions whether we should be optimizing in the first place.

The template of solving for, or letting a market solve for, has been dominant in America for as long as any of us can remember. We're wont to see truth as emerging from a carefully tuned process. In many cases, this means the market, but it can also mean the scientific process, the legal system, the algorithm, the standard operating procedure, or simply a well-reasoned decision. The problem with this, and the source of our modern malaise, is not just the uncertainty. It's also the disconnect we feel between what this calibrated process has promised and what it's delivered.

Depending on whom you ask, we're living in an age of anxiety, an era of narcissism, the fourth turning, or the decline of empire. It's the end of the neoliberal order, of growth; the beginning of authoritarianism, the prelude to dark days or climate catastrophe. A doomsday feeling is rising. *Forbes* magazine claims that the popularity of "pre-apocalypse" TV shows is on the rise, especially among the millennial generation. Optimization has eaten up our time and our attention and even our future, it's filled our bellies and our schedules, and it's left a vacuum when it comes to knowing why.

What motivates this book is a quest to understand how this way of seeing arose and why it took over. How did optimization make inroads? Where has it succeeded? Why has it failed us? And what comes next?

Beginning to understand these questions meant getting out, driving around, and talking to people. I found the stark, lonely farmers eking out the northern plains; the sun-pink shadows of the southwestern desert, hiding nuclear waste and castaway jets; the humid resignation of the Kentucky hills, where horse farms unfurl into an endless carpet of coal plants and distribution centers.

There's something uniquely American about how efficiency became our dominant metaphor. Its paradoxes coincide with our paradoxes as a nation. It grew up in our early churches, in factory lines and computers, on our farms and interstates. It was exported to the world—refined overseas in Six Sigma practices for more efficient manufacturing, in hydropower projects and record-breaking skyscrapers—but it was born and took root on this continent. As the urban-theory pioneer Jane Jacobs writes in her 2004 book, *Dark Age Ahead*: "North Americans prize efficiency. The folk heroes of efficiency of scale—Eli Whitney, Henry Ford, and legions of efficiency experts in their wakes—long ago convinced North American politicians and populations that economies of scale were responsible for America's high standards of living, which is indeed a partial truth."

My goal was to seek out these partial truths and folk heroes, but not under the terms and conditions of optimization—that is, I wasn't interested in a thick tome of dry facts and figures and charts and comprehensive surveys. This is neither a textbook nor a policy treatise. Instead, I wanted to trace how a mathematical concept has taken root in our culture, as told through a handful of characters, each of whom would lend a perspective, a new facet. Like the young man who preludes John Dos Passos's novel *The 42nd Parallel*, I was looking for a snapshot of America at a moment in time, its personas, its sights and sounds, sharing Dos Passos's young man's feeling that "U.S.A." could be found in "the speech of its people." Moreover, I believed this story had a shape, but it wasn't the kind of thing I could conjure out of thin air like an equation. Just as a mathematical modeler oscillates between seeing patterns in the data and constructing the abstract scaffolding to support it, I needed to meet the people who would give flesh to this skeleton idea.

So I got on the road. As I set off, in search of these voices and modern-day shrines to efficiency, I felt like someone studying a religion on the decline. Our monuments of optimization, I came to learn, are often abstract and invisible, hidden behind tall security gates and barbed wire.

And something else surprised me. The people I met didn't just fill in the gaps of the story I'd started. They changed it, and they changed my way of looking at it. A Texas oil magnate revealed how much sadness can accompany even a successful optimization. From a Shoshone ecologist working to restore wild bison, I learned that unwinding an optimization is harder than it might appear. On the Dakota plains, I met farmers and migrant workers both subject to and pushing back against big optimizations in agriculture. From the tidying guru Marie Kondo to railroad tycoon James Jerome Hill, from the wunderkinder of Silicon Valley to a Texas songwriter known to his fans as the Last Knife Fighter, from a Polish-American physicist working on the Manhattan Project to an island baker, I glimpsed the lives of those touched in one way or another by optimization.

The stories that follow are distillations, in some cases, of many years of correspondence and conversation, and, in others, of magazine interviews and television shows and autobiographies. At their heart is a paradox. Optimization was invented, over time, with the idea of making a system, or a society, more perfect. And yet, with a world steeped in optimization, we often see the opposite happening. The factory line was billed as a way to improve efficiency in production, making more things, as well as more leisure, accessible to more people; instead, we have a less equal distribution of wealth than ever. Twenty years of optimizing online search algorithms was supposed to help surface lesser-known ideas and products; in reality, the best-known ones get more and more attention. And this is true for optimization itself, as a way of understanding the world. The more we optimize, the more entrenched we become in that very viewpoint of optimization and the less we're able to do, and see, things any other way.

When I set off in search of the stories that would compose this book, I thought I needed only to find the most representative. The quintessential. The best. That it was all a matter of fitting messy lives into the simple structure of a story. I thought, also, that I was starting from a blank slate.

Along the way, I learned that these stories didn't exist in a vacuum but were interwoven, in ways I couldn't have predicted from the get-go. I learned as well that, like the ruins upon which the Amazon bulldozer was building, my own reliance on optimization ran much deeper, even and especially as its tenets ceased to hold.

The book that follows is roughly divided into three parts. The first considers how optimization functions, how it came to dominate our perspective, and what we've lost as a result. The second explores how the idea of optimization is wearing thin. Just as an overoptimized system eventually breaks, optimization as a metaphor is itself breaking down. Two of the most common responses to this breakdown are to seek more optimization in order to solve things and to seek to dismantle the perspective completely. We consider why each of these is appealing and why each can also fall short. The third part of the book takes a stab at what's next. I believe that the way forward is neither a rejection of optimization nor a doubling-down, but a choice to look past the metaphor at hand.

Half a lifetime ago, my sister said to me, exasperated, "Stop being such an optimizer."

We were walking around a small town somewhere, one we didn't know well, with a few hours to kill and a couple stops to make: find a grocery store, a hardware shop, a place for lunch. It was before smartphones, and we had no map. I began plotting aloud possible paths to maximize the chances we'd hit all of our spots in as little time as possible. She then told me to shut up, slow down, and relax.

She was right. It was a sunny day, and we really had nothing to do beyond enjoying the weather and walk and conversation. There was no need for me to optimize.

Yet her entreaty left me wondering: What is the alternative to optimizing? It surely wasn't picking the least optimal or the worst route. Nor was it complete randomness. Clearly, we had some sense of direction and would eventually stop somewhere for lunch, passing by a place if it looked really bad, going in if we were hungry and it looked good. Nor was it some algorithmic intermediate such as searching for x number of minutes and then choosing the next best place.

The alternative was neither to embrace more optimization nor to rebel against it. It was, quite simply, to drop the language of efficiency entirely.

1

FLATLAND REVISITED

TWO MEN STAND BESIDE A STEEL-CLAD BARN IN GRAND FORKS, North Dakota. The taller of them, Bob, is a farmer with a slight stoop. The short stocky one is a roving worker named Roger.

It's a sparse early April in 2015. Several miles beyond, a factory belches out steam. Sugar beets from last year's harvest are piled up on the ground, preserved by the cold winter. Inside the factories, a byzantine set of conveyor belts and furnaces refine beets into sugar. The refined powder is funneled into bags. The bags are boxed, palletized, and loaded onto railroad cars to be carted out across the continent.

A generation ago, fewer than ten trains came through Grand Forks each day. Now it's more than forty. Some trains carry oil or workers from the Bakken fields out west. The rest carry sugar. Like oil, sugar is a protected commodity and one of the best-funded lobbies in Washington, DC.

Bob's never been to Washington. But he's always liked history. Sugar and national security, he likes to say, are kissing cousins. Napoleon ordered hundreds of sugar factories built in France after a series of port blockages threatened supply. Controlling supply meant controlling a nation. Today, the US government mandates that 80 percent of sugar be grown domestically and guarantees a price.

Of that 80 percent, about half comes from beets grown in the Dakotan plains. The other half is from Florida- and Louisiana-grown cane, its tall canopies flourishing in the thick southern heat. Frigid plains and humid swampland: two worlds apart. In the northeast corner of North Dakota, the Red River Valley stretches out like an ocean, languid white through the long winter, soft beige when the snow finally melts. Cultivated fields turn kelly green at summer's peak. Roads bisect the plane like infinite crosshatch; a few red barns stand out against the skyline.

Bob's family has been farming this land for six generations. His grandfather's tractor is parked in the old barn, and Bob can recount who sold what land to whom and when, like a baseball fan reciting his team's stats. He's lived through lean times, and he's learned to make do. But the future he imagines, from his tiny office overlooking the fields, unfurls with the same sweet cadence as the past.

It feels impossible to imagine, on this crisp April morning, that over the next year Bob's way of life is due to change in big ways.

Bob lumbers to the warehouse that shelters his farm equipment in winter. Later, we see his weathered face bowed in a stern Lutheran service, pulsing with recitations of sin and salvation. The story is well-worn: Two generations ago, half the congregation worked the land. Now he's the only farmer among the nearly full pews. One by one, the younger generation moved to Sioux City or Minneapolis or even farther. With the oil boom, boys went west for weeks at a time. The money there was good. Parents with no children to take over sold off their struggling farms. Maybe a son or daughter would move back eventually to raise a few hogs, plant a few rows, and work at an office in town, but their children certainly weren't going to stick around Grand Forks if they had the choice. Most of the heads in church, like Bob's, are crowned with white hair or none at all.

Farming, of course, is more than an occupation. It's a connection to land, a thread across generations, a means of carrying on tradition. Bob is consumed with figuring out the succession plan for his operation. His face furrows when he reveals that none of his children farm. He later admits, quietly, that he had to cut back grand plans for his farm to take care of his family.

The pastor asks the congregation to stand. In 2015, sugar beets are a lucrative and rooted industry, but it's an existence predicated on the strength of the DC lobby, on the strength of a few families to work the land.

The second man by the barn, Roger, is a laid-back storyteller who hasn't stayed put in years. While others of his age are pinned to one place, for ten years Roger and his wife, Charlotte, have traveled the United States by RV in search of work. In North Dakota in the fall, they harvest beets for Bob. Sometimes they return in the spring, to help process the remainder before planting starts. They've also worked as guards at a Texas oil well, as hay balers in Kansas, and on feedlots and ranches all over the west. As teenagers they baled hay for two cents a bale.

Roger describes the first time he met Charlotte: "There was the prettiest girl I'd ever seen on her family's farmhouse steps. Behind her I could see acres of corn and three oil wells pumping, so I thought not only am I going to marry this girl but I'm going to be rich. Boy was I wrong about the second part." Turns out the farmland and oil rights weren't worth much.

Come October, the couple winds their way back to North Dakota. Much of the state's young male workforce is still eking out a living from Bakken oil. While the earlier boomtown wages have dropped considerably, it's still much more than they could make in Grand Forks. Now the beet harvest increasingly relies on the labor of retirees like Roger, often living out of their RVs. As Roger drives, the radio interrupts his conversation: a trade dispute with Mexico. The big food processors consider sourcing more cane from Florida.

Roger and Bob are cut from different cloths. A stern farmer and a merry nomad. But both of their lives are shaped in major ways by the twentieth-century idea of optimization.

WHAT IS *OPTIMIZATION*? THE WORD CONJURES MATHEMATICAL equations and factory lines, but its etymological roots are somewhat simpler. The Latin *optimus* or *optimum*, meaning the best, made way in the eighteenth century for the French *optimisme*, or the tendency to see the best in a situation. The word drew from the philosopher Gottfried Leibniz's ideas about the goodness of God and soon rushed into the popular vernacular via a somewhat scandalous novella.

Voltaire's *Candide* was published anonymously in 1759 and immediately banned for blasphemy (the author's identity soon came to light). The story follows Candide, a hapless man who experiences tragedy after tragedy—many of them based on true events, such as Lisbon's 1755 earthquake—alongside a bevy of associates, including his ridiculous mentor, Professor Pangloss. At every turn of bad fate, the professor announces that things cannot possibly be as bleak as Candide would have, because the two of them are living in the best of all possible worlds. Pangloss, in other words, espouses a philosophy of l'optimisme. The term *Panglossian* has come to refer to one who partakes in overly optimistic buffoonery.

Pangloss, as readers of the time surely knew, was modeled on Leibniz. The philosopher had put forth a somewhat circular argument, claiming that God's goodness resulted in proof that any world of his creation was the best one that could possibly exist. In laughing at Leibniz's reasoning, Voltaire sets up a useful definition for modern-day optimization.

Suppose you set out to start a lemonade stand that serves the best possible lemonade. How would you go about it? A good first step would be to define what constitutes the best, anyway. You may decide the best lemonade is the sweetest, the tartest, or some proportional combination of sweet and tart. Perhaps you want to stay below a given percentage of sugar or above a certain level of acidity.

You might also look in your pantry and see what's there. Only two cups of sugar? Maybe you'll need to make less lemonade. Do you have ten pounds of lemons to use up? Can you borrow some corn syrup? Say you've got a hundred thirsty customers lined up outside on a stifling hot day. What would the best lemonade look like now? Maybe it's a little tangier or more watered down.

These are the questions at the heart of an optimization. They suggest the basic framework behind modern software used in maximizing everything from grocery inventory to airline schedules. First, you define your ideal lemonade, or objective function. Next, you see what you have to work with, or your parameters. Finally, you consider your constraints. Things get a bit more complicated when there's dynamism in the system, or nonlinear relationships, or when you have more than one objective or the dataset is very large. But the same underlying principles apply.

The process of finding the best lemonade is the optimization itself, which is carried out using one of many techniques developed over the past few centuries. Solving for the best lemonade might be as simple as writing out a recipe: 2 lemons + ¼ cup sugar + 4 cups water. Or it might be as complicated as a computer running millions of calculations to iteratively find the best results based on the data points—local prices, taste tests, global lemon-supply projections—you've identified.

Let's say your lemonade stand is successful, the thirst of your first hundred customers quenched. You're feeling good about yourself, but you start to wonder. Is yours truly the best of all possible lemonades? How do you define all possible lemonades, anyway? Does the scope of lemonade allow for other ingredients, such as lime or raspberry? Do you compare your product to that produced in countries where lemons are abundant and inexpensive or have an otherworldly fragrance? If your lemons are fifty dollars each, does that change your equation, or are you insensitive to price?

All of these considerations reveal important facets of an optimization. How do you bound your function, and how do you know when it's complete? When do you update your parameters and constraints to solve anew? How well does your particular optimization generalize to other regions, other situations? To what extent can you plan ahead and cement your optimized process in place? Or are things changing so quickly that there's no point in nailing it down?

The lemonade example also sheds light on why optimization might work better in some contexts than others. If the price of sugar and lemons is stable, if you own a lemon grove and have an abundant source of clean water, if your customer base has had the same taste in lemonade for a thousand years and you don't expect things to change, then a large operation built around a fine-tuned lemonade recipe makes a lot of sense. If conditions are more fluid, you might prefer a small lemonade stand to a large, locked-in approach. On the other hand, being large might make it easier to scale up production.

This simple process of optimization has been refined over the past few centuries, and formal mathematics has grown up around it. In recent decades, advancements in computing have allowed for new orders of magnitude of calculations, testing contingencies, and counterfactuals. The same core idea, this search for the best of all possible worlds, has been applied far and wide, from scheduling cargo ships to auctioning off digital ads, from the development of new pharmaceutical medicines to the production of cars. And it's at the heart of how American agriculture went big.

HUMANS HAVE BEEN TINKERING WITH CROPS FOR MILLENNIA. FARMers, and later, biologists, have sought systematic ways to squeeze more nutrition from less land and labor. But it wasn't until the twentieth century that the push to optimize became a powerful force in farming, thanks in part to a Midwestern farm-boy named Norman Borlaug. Born in Iowa in 1914, the grandson of Norwegian immigrants, Borlaug opted to leave the family farm, seeking a college education and travel abroad. An Iowa congressman later attributed the young man's drive, perhaps apocryphally, to the advice of Borlaug's grandfather: "You're wiser to fill your head now if you want to fill your belly later."

Borlaug took these words to heart, traveling far from Iowa to fill his head, and he eventually completed undergraduate, master's, and doctoral degrees in forestry and plant biology. Borlaug's "Green Revolution" spread key innovations in modern agriculture around the world. By breeding for dwarf varietals, he engineered a wheat less likely to fall down with weather, as well as easier for machines, but not humans, to harvest. His idea of backcrossing—or repeatedly breeding a hybrid line with a parent plant—increased disease resistance substantially.

Borlaug's natural charm and agronomic successes attracted many disciples. In a single generation, their research increased crop productivity severalfold. Borlaug is credited with keeping a billion people out of starvation. In an *Atlantic Monthly* tribute, science writer Gregg Easterbrook claims the agronomist "saved more lives than any other person who has ever lived." Borlaug garnered these kinds of superlative accolades left and right, among them the Nobel Peace Prize, and his work profoundly influenced not just academic research but also the more commercially minded endeavors of agricultural-technology corporations.

But it was Borlaug's way of seeing nature as something to be harnessed that would forever shape not just agriculture but our entire modern world.

Just north of Borlaug's birthplace, cooperatives of beet farmers like Bob have been cultivating for centuries the belt of prairie that stretches from Montana and Idaho in the west, touching northern Colorado, then east through the Dakotas into Minnesota. The bulk of America's sugar-beet crop is grown in the Red River Valley of North Dakota, about an hour north of Fargo.

The co-ops there maintain community governance, and family farms still own the land. For many generations, not much changed. Then slowly, in the late twentieth century, farms began to consolidate. This was the result of a few major trends. First, the revolution in plant breeding had given rise to a whole new set of seeds and technologies, all of which allowed fewer farmers to operate more land. Others were freed up to seek more lucrative jobs in town, which many had already sought because of a second factor: a decline in farm incomes. Innovations in commodity markets changed who bought a farmer's crop, how, and for how much. Speculation created uncertainty. Land prices rose as revenues fell.

As a result, operations were squeezed from both ends. Higher fixed costs and lower grain prices meant they had to get big or fold. Many in the Red River Valley gave up the traditional four-crop rotation in favor of growing beets year on year. Crop rotation allows soil nutrients drained by one crop to be replenished by another. It also spreads the farmer's risk across multiple foods: if weather wipes out beans, the wheat may yet make it. By contrast, a mono-cropped farm invites more pests and weeds and, ultimately, risk.

As farming intensified, the arms race with weeds accelerated, and as it did, seed companies sought newer and better varietals, with more resistance to disease. They hired teams of plant geneticists and chemists. These scientists needed to be paid nice salaries, so the companies charged more for seed. Farmers, now managing tenfold more acres than a generation ago, still couldn't keep up with weeds on their own, as they had in the past. They hired agronomists and advisers, and sprayed on herbicides wholesale. In parallel, chemical companies saw enormous opportunity in the excess ammonia developed for World War II munitions, just as land-grant universities were spearheading research around how specific nutrients interact with crop growth. The results were new classes of chemical fertilizers. Yields per acre went up.

Ezra Taft Benson, the secretary of agriculture in the 1950s, proclaimed infamously to farmers: get big or get out. This guidance would come to feel ever more oracular as the century wore on.

Greater yields meant farmers had more money but less time. They began to invest in fancier equipment. Crops were bred for mechanized harvesters, and harvesters built for modern crops. Their advisers told them what seed to purchase; their mechanics maintained equipment. Bit by bit, the grit under a Midwestern farmer's fingernails disappeared, his skin paled. He was spending more time in the office, selecting the best varietals and thinking about futures contracts, than in the field.

"I'm a person who wears a lot of hats," Bob says. "Sometimes a few are on fire."

As for the seed companies' scientists, they were carrying on Borlaug's revolution, optimizing for varietals that would yield more. Beginning in the 1980s and 1990s, traditional breeding efforts were augmented with the splicing in of genetic code to resist drought and frost and pests. Down the hall, other scientists worked on chemicals that would combat weeds without killing the crop. These scientists knew, as did their business teams, that farmers would pay nicely for seeds that lessened their work. It was easy to ignore, for a while, the costs: the chemicals harming human health, the machinery depleting soil, the fertilizer spewing into the downstream water supply.

Just as the scientists and farmers like Bob lost track, we did, too. The irony is that optimization created a language for tracking trade-offs precisely: tenfold more expensive machinery meant ten times less time harvesting. Organic varietals cost more and required more time but could be sold for a higher price. It was all worked out in a spreadsheet. Even if the seed-company scientists weren't measuring the environmental trade-offs of their products, the costs were readily visible downstream.

Every gain, every trade-off could be quantified. What we lost, however, was the language to describe the trade-offs of optimization itself.

VOLTAIRE'S SATIRIC BEST-OF-ALL-POSSIBLE WORLDS WAS APPLIED retrospectively by Professor Pangloss to explain away the various tragedies that befell his disciple Candide. By contrast, when we optimize, we are predictively looking forward, imagining all possible worlds and carefully assessing them to get to the best. How can we grow more sugar, per acre, for less money? By how much would buying a new quarter-million-dollar tractor lower labor costs? What is the right combination of ingredients to make the best lemonade, at scale?

Scale. More. Better. Faster. These words sound familiar because they dominate our language, not only in business but in our daily lives. Every year, our cell-phone contracts promise yet more data and faster speeds. Gym chains and diet plans offer better results, sooner. A small-town ice-cream entrepreneur is asked the inevitable: "But how does it scale?" Optimization has become the lens through which we, in the West, see the world. This lens has been built on a set of principles and assumptions, and it has resulted in a handful of major changes over the past three centuries.

One of these changes is a material change. Increasingly, we see the world as broken down and specialized, into hours of the day, job functions, and levels. We talk about our competitive advantage, discuss performance, and track productivity. We eke out ways to make our diets and our childcare routines more efficient.

Optimization has also changed our point of focus, from one of observation to one of control. Increasingly, we seek to harness nature, to engineer a world around us. If it can't be measured, it can't be managed, or so the saying goes. And the goal of management is to improve, or optimize, a system or society.

Perhaps most of all, optimization has changed how we understand the world. When we become good at controlling nature with our ever more miraculous inventions, we start to assume we can solve anything. We see a destiny that grows ever upward, progresses ever forward, toward something higher. We embrace the belief that growth is good, that the world we've engineered is the best that could possibly be.

Nowhere has this kind of momentum been more pronounced than in twentieth-century American agriculture. As historian Paul Conkin summarizes in his book *A Revolution Down on the Farm*, productivity growth in agriculture has been nearly triple that of other industries: "Since 1950, labor productivity per hour of work in the non-farm sectors has increased 2.5-fold; in agriculture, 7-fold." Moreover, "In one generation from 1950 to 1970, the workforce in agriculture declined by roughly half, where the value of the total product increased by approximately 40%." In other words, the most dramatic contribution to efficiency has been brought forth by the industry that produces the very food we eat.

Voltaire famously ends his novella with a pompous Pangloss declaring that all their tragedies were in fact for the best and the hesitant Candide replying, "We must cultivate our garden." Some see Candide's statement as the high-water mark of sneering nihilism; others consider it a romantic retreat into nature and the incremental work represented by gardening.

For a farmer whose crop touches global commodity markets, Bob may be as close as it comes to an old-school way of farming. Dakotan soils aren't as rich as those in the lower Midwest, and the growing season is short. It can freeze into April or May, and the temperatures plunge again in October. Bob, like just a handful of his neighbors in the Dakotas, grows not only beets but also dry beans and wheat.

Yet Bob stands apart in one major way: his ongoing resistance to GMO. Over the past few years, the rest of the Red River Valley slowly shifted to genetically modified seed, while Bob alone held out. This choice, it turns out, would be critical not just to the future of Bob's farm but to that of the entire region.

The story of optimization drew me first to North Dakota, because agriculture lies at the heart of optimization's project, and because I'm drawn to stark landscapes. I see them as fertile ground for new ideas. Indeed, the Dakotas are among the forgotten places of North America. At a museum in Fargo, a bustling town with brick bistros and art galleries built into now-shut-down factories, I stop in to see a photo exhibit about the Bakken oil fields. It's filled with black-and-white pictures of large machinery, burning plumes lighting up night skies, and workers in thick parkas set against a snowy canvas.

I had notions of making a short film about Bob and his family, a place and a way of life in slow decline. For all the sugar packets that grace diners across America, few know of the Red River Valley or where the sugar they consume is made. I spent time with farmers all over the valley, talked to leaders of American Crystal Sugar, the co-op behind the area's farms and factories, and its lobby arm in Washington, DC. I visited the cane plantations in Florida, attended church services, and met Bob's congregation. Bob let me into his world, and his struggles, with unflinching generosity. For my part, I remained curious but reserved, letting on little about my personal life.

I thought I was the one asking questions. It was only after a church service, when Bob sent me an email to inquire as to whether I was interested in joining the flock, that I realized perhaps I wasn't the only one on a mission.

FOR ALL OF ITS SUCCESSES, ITS BILLION LIVES SAVED, THE GREEN Revolution was not without cost. In many countries, more-focused crop breeding made the difference between starvation and a well-fed population. In America, however, the deification of efficiency cast aside traditional methods of letting soil rest and regenerate. Its focus on increasing yields divorced agriculture from farmers, and farmers from land and place. And perhaps most perniciously of all, as the buyers of crops became accustomed to higher yields at lower prices, a whole system melded itself around those customs.

Bakeries grew from mom-and-pop operations into specialized factories. They no longer sold bread and pastries to a single neighborhood, but to the world at large. This required new transportation arrangements, new negotiations with grocery chains, new advertising to consumers. Consumer appetites changed, too. With food production specialized, prices dropped. Fast food became possible. Routines changed. With a meal at the local McDonald's so affordable, there was no need to pack a lunch from home. There was no need for an office to staff a cafeteria. We all know how the story goes. As big-box stores and fast-food franchises proliferated, local shops floundered. It was hard to go back.

Throughout the second half of the twentieth century and the early twenty-first, the sugar-beet co-ops of North Dakota got used to the slow, steady uptick in technology, from bigger tractors to higher-yielding seed. Change became a constant. But it was still possible to keep track of things. Updates could be delineated by generations: Dad did it this way. When I used to ride in the tractor with Grandpa, he did it that way. Nostalgia mingled with the comforts of a bigger house, a college fund for the kids. The pace of change could be jarring, but it still felt in keeping with life's rhythms.

Then, suddenly, a lot changed, and fast. In mid-2015, the major sugar co-ops voted to grow only GMO beets. This meant growing their crop from seed that had been genetically modified to accept the herbicide Roundup.

This choice came after several years of debate. Over the past seasons, farmers had been contending with the threat of price fluctuations, due in part to Mexican producers' "dumping" sugar on the market. The argument for GMO was one for greater certainty, more streamlined operations. Instead of spraying chemicals every week, as they had with non-GMO seeds, farmers could spray Roundup once per season. Yes, they'd be locked into the seed company's prices and more limited GMO selection. Yes, they'd lose some of the old practices and some diversity in the seed. But they could continue to increase yields, with less labor.

Many co-op members were on board with the choice to go GMO. They begrudged the seed companies their monopoly practices, sure, but GMO felt like the future. Hadn't farmers been breeding better crops for centuries? The new seed meant fewer chemicals. It felt like a step in the right direction.

A handful questioned the decision, and then there was one holdout. Bob saw a different future, or perhaps he felt wedded to the past. To him, GMO represented an important step change, a break from history. It felt unknown and unmanageable, compared to the kind of farming he'd learned from his father and his grandfather. He wasn't sure about Roundup. He liked making his own decisions about what to plant where and how to manage his fields. He didn't like that everyone would be growing the same thing. He harbored a deep mistrust of the seed company's iron fist. He wanted his autonomy. He even enjoyed spraying once a week—it felt like forward motion.

Consumers today are familiar with the GMO question from a different angle: that of consumer health. Studies have shown Roundup accumulates to cause deleterious effects in our bodies. In 2021, Bayer was ordered to pay upward of $10 billion to settle claims linking the herbicide to cancer. But in 2015 in North Dakota, these concerns were far from a farmer's mind. The DC lobby arm of the sugar co-op started showing up at shop meetings, helping farmers set up blogs and Twitter accounts to express excitement over the agronomic step forward. A couple of younger members took to the technologies, but most had altogether different concerns. In community meetings, one question came to the fore: whether their farms, and their way of life, would survive.

The answers from the co-op management fell out along clear political lines, although these lines had long meandered and blurred. The co-op farmers generally wanted the federal government to stay out of their business, but they also wanted federal subsidies. They resented the seed companies but also relied on them. They laughed at "city folks" who couldn't tell a cornstalk from a mop handle, yet these were their customers.

The lines blurred especially when it came to what GMO meant for agriculture at large. Every farmer I've spoken with feels a connection to his or her own land that far transcends the economic. Many saw the potential of GMO seed as a way of being easier on the land. For his part, Bob could admit that maybe there were shortcomings in his current practice. It wasn't organic. It wasn't how the land was meant to be cultivated. There was something spooky about weekly sprayings. But GMO wasn't the way out. His father and grandfather hadn't needed GMO. They would've scoffed at today's giant tractors and electronic farm records. They were in their rigs with pencil and paper, calculating and planting and spraying and fertilizing and harvesting from April to October. Just as Bob intended to be.

What happens when we optimize? We find a solution. We scale it. Things work. And we grow accustomed to the fruits of the scaled invention. This was as true of the new GMO seed as it was of Bob's way of doing things, which itself had been optimized over centuries, first by selective breeding and then by synthetic fertilizers and larger equipment. It was hard to get off the treadmill.

The year after the co-op vote, beet-harvest numbers skyrocketed. So did revenues. The seed was performing as promised. It was good to have everyone on board, doing the same thing. American Crystal management was over the moon. Seed sales reps patted themselves on the back and charted how they'd expand their reach to other parts of the plains. They sketched scenarios: Should they raise prices slowly this year, when spirits were high, or sneak it in next season?

When the next season arrived, something unexpected started to happen. It surely wasn't something the farmers in Grand Forks, many of whom had traveled only once or twice, if at all, to places like New York or San Francisco, saw coming. There was a sudden sea change in consumer attitudes. Concerns once relegated to health nuts and coastal environmentalists suddenly hit mainstream. The vote was in: GMO is bad. GMO is dangerous. Grocery shoppers, not just on the coasts but inland, too, looked for GMO-free labels. Major food producers, beginning with names like Hershey's, announced plans to cut GM ingredients entirely. And nearly overnight, food manufacturers sourcing sugar turned their eyes southward, to non-GM cane grown on Florida plantations. The agronomy of cane hadn't lent itself as easily to genetic modification (although the research eventually caught up, and GM cane is now being grown in several countries outside the United States), so panhandle farmers had never had to face the choice.

The Dakota factories sputtered. Beets piled up on the ground outside. Sure, 2016 wasn't the worst of harvests, and prices were still holding up all right. But the writing was on the wall. Unless they changed, and fast, this was the beginning of the end, not just for the farming tradition of the Red River Valley but for their way of life and the livelihoods of migrant workers, such as Roger, for their churches and coffee shops and adult children who'd surely never move back home from the cities now. Bob felt a certain smugness. He'd been right after all.

But had he? He may have opted out of optimization, but he'd nonetheless go down with the ship. Because of the co-ops' hold, he had no market, no lines of distribution for his non-GMO beets. The sucrose in beets degrades rapidly, so the crop can't be stored or shipped as grains might. At most, beets will last a few winter months, preserved in frozen piles outside the factories. Which means the production must be local, and the Red River Valley's factories weren't going to process Bob's paltry crop separately, without other non-GM growers on board. And he still had no succession plan for his sixth-generation farm.

MOST OF US AREN'T FARMERS, BUT WE'VE CONSUMED THE FRUITS OF Bob's labor. We mix sugar into our morning coffee. It's in our muffins and breads and cookies and cakes, in our meat and our vegetables, too. It's programmed into the value of the stocks we trade and the taxes we pay, into our health care. The omnipresent set of sugar packets in coffee shops and on diner tables, from Alaska to Maine, is the result of layer upon layer of optimization.

Beets and cane are grown on ever-larger operations to maximize yield per acre, then harvested by larger and more automated equipment, and processed in factories that churn nonstop, in order to optimize the amount of human and material energy needed to create a unit of food. Transportation has been optimized to get crops to producers in record time. This involves not just building out roads and railway lines but scheduling fleets and drivers along the fastest, cheapest routes. Grocers and restaurants can order supplies as they need them, rather than in advance, to minimize the time items spend on their shelves or in their pantries.

Markets run in near-real time, and the price of commodities reacts to changes in weather forecasts or consumer sentiment. A baker no longer has to buy all his supplies at harvest, only to risk degradation or theft. And a farmer no longer has to sell his crop at the same time, when the market is flooded and prices are low.

You might call some of the trading less than efficient. The lobbying in DC certainty isn't optimal, and you might have your opinions on whether tariffs are good or bad. But the overall effect is clear: in 2020, adjusted for inflation, consumers paid a third of what they did a century earlier for sugar. Food is cheaper and more abundant in twenty-first-century America than at any time in human history.

These efficiencies have come with trade-offs. Borlaug may have saved his billion lives, as large swaths of the world moved from subsistence farming to reliable supplies of calorie- and nutrient-packed grain. At the same time, studies have established links between industrialized food systems and chronic disease. In the eighteenth and nineteenth centuries, it was crop rotation that allowed agrarian societies to produce more healthy and abundant crops, which in turn allowed more people to move into urban areas and take the jobs created by new industries and factories. Mid-Victorian Britain was perhaps one of the healthiest societies in history, surpassing our own in terms of nutrition and health outcomes. In the decades that followed, however, outcomes went off a cliff. Average height dropped by a couple of inches. Historians attribute the decline in health to the very things that, a generation earlier, had enriched them in the first place: industrialization, rising wealth, the move to urban centers.

Soil and environmental health have suffered as well. One study finds that the American Midwest has lost over fifty billion tons of topsoil since cultivation began a century and a half ago. The nitrogen and phosphorus fertilizers running off fields into waterways have created massive dead zones in the Gulf of Mexico. And while the farmers who have survived may enjoy some of the fruits of consolidation, the quiet office in place of the scorching field, others see communities once held together by agronomy eroding.

Some of the trade-offs are epistemic. As we saw when the Green Revolution came to North Dakota, and as we'll see in the coming chapters, the very act of optimizing entrenches a perspective. In fact, the idea of trade-offs is one borne of this mindset. It flattens the complexities of agriculture into a formula; it turns a field of beets into a blank canvas to be engineered; it cements the idea of forward progress and trading tit for tat.

The result is a world that may look familiar to one of several generations ago—the flat plains are still there, the barns, the roads—but much diminished. The property lines that once hemmed in a family's history have faded, written over with the records of crop yields. The fields look flatter, spare, hollow, although they're producing more beets than ever. With efficiency, we gain in productivity but suffer other losses.

We can think of these losses as the flip side of what we've gained from optimizing. We'll call them, in brief: slack, place, and scale. Slack, or redundancy, cushions a system from outside shock. Place, or specific knowledge, distinguishes a farm and creates the diversity of practice that, ultimately, allows for both its evolution and preservation. And a sense of scale affords a connection between part and whole, between a farmer and the population his crop feeds.

The first thing lost, slack, is one often noted in studies of breakdown: of cultures, of organisms, of schedules and systems. As we optimize, there's a necessary compromise between progress and robustness, between speed and the ability to adjust to setbacks. The more tightly wrought our systems get, the more razor thin our margins, the more fragile we become. This has been described by anthropologists who see civilizations collapse on the heels of increasing fragility. It can be seen in biological organisms, which sustain greater injuries and illness when already stressed, and in engineered systems, such as bridges and supply chains, that break down when tolerances are too thin.

As American agriculture eliminated redundancies in seed types and farms, it lost its connection to land and the specific knowledge that comes with it, outsourcing cultivation to the algorithmic demands of increasingly globalized trade. As farms sent their crops abroad rather than to their local community, as family operations were scooped up by farm managers and corporations, slowly but surely this sense of place disintegrated. Today, much of North America's farmland is no longer owned by individual families but by holding or management companies. The richness of a small-town community—its generations attending the same schools, diners, churches, local theater performances—thins and dulls. Stories shared in the coffee shop fade into evenings in front of the TV, watching national and international news. Neighbors no longer stop by to shoot the breeze or lend a hand.

The third thing lost, scale, is perhaps the most subtle and the most critical. The more the co-ops bought into efficiency, the harder it became to opt out. With the first improved seed, farmers lost a sliver of autonomy. The corporations promised more for less. New seeds allowed farmers to farm faster. They could buy more acres. Meanwhile crop prices were going down, and with seed companies and food producers taking a larger slice of the profits, farmers had to farm more acres to break even. Revenues became the metric that mattered. The co-ops cast aside other, less quantifiable goals, such as environmental health and the weave of the community. They equated yield numbers with what was good for the farm. For their families. For their towns. Suddenly the short term was out of sync with the long.

The difference between small and large operations is more than a matter of degree. A small vegetable farm, with greenhouse crops, is more adaptive: new varietals can be switched in as consumer tastes change. A large farm scales non-linearly, and so does its entrenchment. When the beet co-ops voted to go GMO, they built on prior decisions that would take years if not centuries to reverse. And their decision had the effect of homogenizing production, leaving outside individuals, such as Bob, unable to compete. Bob couldn't simply go make his own lemonade in another jug, just as many of the castes and cultures given the short end of the globalization stick cannot simply "opt out" of the optimized system for one that preserves slower, more local, and less quantifiable aspects of their lives.

Over the course of a brief century, American farmers traded the richness of slack, place, and scale for optimization along a single axis: yield. We grew our economy and population and helped engineer advances around the world, but not without some loss.

You and I, our whole Western world, whether we know it or not— we've all made this bargain, too.

FARMERS ARE NOT UNAWARE OF THESE BARGAINS. INCREASINGLY, THE largest operations are taking issues like soil regeneration, water quality, and natural pest management seriously. A farmer I know in Iowa raises some ten thousand acres of organic and conventional corn and soybeans. His is a fifth-generation family operation, and he speaks fervently of the work they've done to preserve the aquifer, to replenish soil sucked of nutrients. A year before I visit North Dakota, I attend a panel discussion he hosts in one of his barns not far from Iowa City. He's also invited a nearby hog farmer and two women who've started a community-supported agriculture project to grow vegetables for the local area. While the political inclinations and vernacular of the panelists differs by a country mile, their attitudes around land stewardship are remarkably in line.

Other farmers point out that compared to many grain-growing regions, the United States is still relatively "underoptimized." In countries such as Brazil, Argentina, and Ukraine, corporate operators might manage a hundred- or thousand-fold the acres of even the largest Midwestern farms. America's preservation of family-run operations is due in part to the way farmland was divvied up in the first place; in part to government incentives since, which have promoted individual farmer-owners over corporate management. Conversely, in yet other countries, Borlaug's revolution has done wonders for small-scale crop breeding, without the dramatic mechanization or environmental consequences seen in the major grain-growing regions of North and South America.

And even in the United States today, it's been shown that a shift to regenerative agriculture can be done without major economic compromises. On an Iowa farm, two systems were compared over twenty years: on one, corn was grown continuously; on the other, a traditional crop rotation without synthetic fertilizers or chemicals. While the first yielded more in gross revenues (because corn offered the best price) and required less labor (because laying down a pesticide is easier than natural pest control), the second netted more, because less was spent on expensive seed and chemicals to keep the pests and weeds at bay. Moreover, the soils improved, such that the farm could continue producing for many more years.

This kind of deceleration was shown to work in an experimental setting, with many upsides and few down, but it hasn't taken off more broadly. Why not? There's an asymmetry to optimization, as we'll begin to see as these chapters unfold. Once the farmers leave the land, once the practices are forgotten, once the tractors get big—it's hard to go back.

Indeed, another Midwestern farmer cites "having girls" as the best thing that happened to his business. While male children are largely expected, and even pushed, to take over the farm, the expectation is relaxed and often abandoned when it comes to daughters. As a result, he neither pressured his girls to continue in his footsteps, nor did he pressure himself to raise the necessary capital to grow the operation, in a time when land prices were going through the roof. This choice had unintended consequences. Instead of meetings with banks and brokers, the farmer's time was freed up to make small but impactful steps toward local efficiencies on his farm, such as manually switching equipment to more sustainable fuel.

Of course, not every experiment in the direction of "deoptimizing" goes so well. A Vermont farm, in the early 2000s, wanted to produce sugar from scratch. The farmers' goal was self-sufficiency, without the heavy hand of co-ops and sugar lobbies and large-scale agriculture. Doing without the heavy hand also meant doing without the investments and optimizations of the industry: from the railroads and processing factories to the lobby groups and the breeding efforts.

They were experienced farmers, and their plans were comprehensive. Nonetheless, the experiment ran into issues. The farmers needed to reinvent an entire process, from planting to refining, that industrial operations had perfected over decades. There were issues with the distillation, producing a taste "best described as over-ripe Swiss chard" with "a chemical-like burn quality."

Erik Andrus, one of the farmers, waxes geopolitical in describing what they learned. He cites France's taking on domestic production during the Napoleonic Wars when Britain cut off sugar from the tropics and concludes that "when circumstances force us" into more local production, farmers will figure out a way. However, "for the time being, in this country, sugar will likely remain an aspect of huge-scale agro-industrial production that we all take for granted."

AT FIRST GLANCE, THE DAKOTA PLAINS SEEM TO STRETCH OUT TO-ward a flat horizon, unbroken as the open ocean. But if on a warm July afternoon you were to pause and examine the lay of the land, you'd soon find nuance. A cultivated field is patched with shades of green and yellow. There's a little hill behind the farmhouse, a thicket of woods to wander into. It's barren by the train tracks. The field across the way is pockmarked with empty beer bottles and the ex-ploratory holes of rodents.

Imagine now that as you sat there with the blue sky above, you could flatten out these plains further, like an infinite piece of putty, erase hillocks and barns and train tracks, mush greens and yellows and reds into a single tone. Imagine that instead of your human body breathing in the skies and fields, you were a simple square, as flat as the surrounding plains. This is the premise of the nineteenth-century novella *Flatland*, by Edwin Abbott Abbott.

Originally penned as commentary on the Victorian social order, *Flatland* is also a mathematical parable. The story features a two-dimensional protagonist, named A. Square, who lives in a two-dimensional world called Flatland. In his travels, Square chances upon a couple of new worlds: Lineland and Pointland.

In Lineland, objects are expressed as projections onto a line. So Lineland's inhabitants see Square as a line. A circle also looks like a line. Triangles are lines. Lines may be longer or shorter, depending on the object's size, but the richness of Flatland's two dimensions is lost.

In Pointland, the view is even more limited. Lines of any length become points. Squares and triangles are no longer distinguished even by the lengths of their sides. Instead, every object is the same.

Square, upon discovering the relative closed-mindedness of Lineland's residents, embarks on a mission to enlighten them. He speaks of the richness of Flatland, describing with great eloquence the difference between triangles and circles and squares. There are hexagons, even! Not surprisingly, the residents look skeptically at this itinerant guru, with his claims of "higher dimensions." After all, his "triangles" and "circles" look pretty much the same to them. Just lines.

It's only when Square returns to Flatland, and welcomes a visitor, that he begins to comprehend his disconnect with Linelandians.

Square's visitor is named Sphere. Sphere looks a lot like Square's friend Circle, and yet there's something different about him. Sphere speaks animatedly of his home, a three-dimensional place inhabited by cubes and pyramids.

Square is suspicious of Sphere, fearing a snake-oil salesman of the worst kind: affable and enthusiastic, a stranger in town, hawking promises of riches just beyond reach.

Sphere, however, is not just affable but relentless. He sees in Square a ready disciple. To convince Square of this third dimension, Sphere pulls a magic trick with his own body. He moves up and down and through the third dimension, intersecting the plane of Flatland, such that to Square, it looks as if Sphere's circular shape is growing and shrinking in size. From a point to a broad circle to a point again, as his friend moves through the plane. Square sees it now. He understands that Sphere is speaking truthfully, albeit a truth framed differently than his own. And he now understands why the residents of Lineland looked at him askance.

Square's framing of the world, his two-dimensional plane, is an optimization. A way of seeing. Many of our own optimizations are just that: metaphors for how we organize the world. When we think about efficiency, we begin to see places where it might be applied.

A model, like a story or metaphor, is a way of bending reality into shape. It in turn shapes how we see the world, freezing itself in place, reinforcing the chosen frame.

By embracing efficiency above all, we've crowded out what can't be measured and optimized and allowed the metaphor of optimization to cannibalize other worldviews. We've fooled ourselves into thinking we can solve the problems of optimization with more of the same. And as a result, we've paradoxically slowed our ability to change.

Things didn't turn out so badly for the Red River Valley farmers. Consumers soon turned their attention from GMO to other health concerns. Some food producers stuck with non-GM cane, but many walked back early promises. After all, as the proponents repeated to me countless times in North Dakota, when you test processed sugar, there's no way to tell the difference between GMO and non-GMO. A molecule is a molecule.

Of course, they never spoke—publicly, at least—of how GMO might be affecting not just the sugar molecule but their agriculture-based livelihood.

I'VE LOST TOUCH WITH BOB. IT'S BEEN SEVERAL YEARS SINCE I VIS-ited North Dakota, since I hung around watching the planters go out in the cold spring and the harvesters roll in the fall, since I toured the beet factories and stood beside the railroad tracks chatting with Bob during his summer lull.

The short film I'd been working on never came to fruition, nor did Bob's attempts to bring me into the religious fold. Our interview footage is gathering dust on a hard drive.

As I look back, I realize I was never really sure what to make of Bob: Was he a martyr, an oracle, a hopeless relic? He wasn't the par-agon of what I believe needs to be set right in agriculture, but in some small way he was resisting the forces of forward momentum. For my part, I was already an outsider in North Dakota, peeking in, and I feel outside anew as I revisit our interviews.

When I call Bob to catch up, his daughter answers the office phone. He calls me back a few hours later on his cell phone, and his first words are, "Well, isn't it a different world, than the one we last saw each other in?"

Indeed, it is. A lot has changed since 2015, and Bob ticks off notes on the past few years: 2019 was a terrible year; they lost a third of the beet crop; 2020 was even worse; they lost half the crop and had to sell off some equipment to break even. They'd braced for the worst in 2021, and it again surprised them with the best yields ever. Bob calls it a "healing year."

I ask him if he's still growing non-GM beets. He sighs, and I know what's coming.

A few summers ago, a European seed company came to field-test seeds on Bob's acreage. The company was testing traditional breeding updates for non-GM varietals. Bob was the only grower in the region still interested in non-GMO. Even though he later learned the Europeans were planning to add in GMO down the line, he was thrilled. He even considered buying their product. But beets are picky; varietals do well only in certain areas. Not even the European company saw value. That's when he realized, "Guess I might as well just fold."

Bob doesn't sound unhappy with his decision so much as resigned, as he reflects on his negotiations with American Crystal. "I just couldn't hold out any longer . . . I tried to press Crystal's hand; they weren't interested. I was just one guy in their way."

In the end, Bob's farm, the five generations behind him and the two already ahead, took precedence. He planted his first GMO crop in 2018. "It's so much easier," he tells me with a hint of chagrin. Turns out his neighbors weren't wrong: "Now it takes one person to farm a thousand acres instead of four. They all wanted to work Monday through Thursday and buy a lake cabin to go to on weekends. Well, they got that with GMO."

In the mathematics of optimization, the No Free Lunch theorem has it that when you reduce the complexity of a model, you also reduce its descriptive powers. In other words, there are no shortcuts, no ways of getting something for nothing. Something important happens when you bound a system. You're forced to consider the trade-offs, the losses alongside the gains, as Bob and other farmers have done. For a long time in America, in a world wide and plentiful, making things more efficient made sense. When the boundaries were closed, optimization became easier. We understood the constraints. At the same time, as in Flatland, our minds closed, too.

In the next two chapters, we consider the question of how we got here. How did we trade the enormous gains optimization has provided, for the loss of slack, place, and scale?

As any farmer knows, lying latent in any story of loss is one of rebirth. Bob found no market for his non-GMO beets. Square returned to Flatland unsuccessful, ostracized for his heresy. But he brought home a new way of seeing. Even when all is lost, the seeds remain, along with the possibility of carrying forth something new.

2

LEAVING LAS VEGAS

"THE LONELIEST ROAD IN AMERICA" IS A TITLE THAT OFFICIALLY BE-longs to Highway 50, which cuts west across the state of Nevada. But when it comes to loneliness, nearby Highway 93 would give the official titleholder a run for its money. Highway 93 wends its way from the grayscale Snake River Valley of southern Idaho, via the sad three-casino-motel town of Jackpot, through scantily populated reservations and high desert no-man's-land. It branches out into Highway 255, paralleling a hundred miles to the west, through the near-abandoned gas station and grocery store in Owyhee, where a display case that hasn't been opened in weeks offers baby shoes and beaded earrings. There's a DVD rental, a pot of warm coffee, the freezer section with three or four microwavable pizzas. The produce feels as gray as the cold air outside.

The wind howls, carrying stray tumbleweeds. A mess of them covers the entire west side of the road for some hundred feet, turning two lanes into one. Cars pause, as if crossing a single-lane bridge, to let another pass. At the junction of 93 and Interstate 80, things liven up a bit. A dozen trucks are wedged in for the night. In the gas station's parking lot, a man empties an ice cooler from his camper van. A lone motorcycle whizzes by.

Highway 93 then skirts beside a couple of ghost towns. It winds, gray on gray, gusts and high plains, until—when you least expect it—the road hits another interstate, where it smash-bang runs up into the glimmering dazzle of Las Vegas, the city's endless horizon and sky-scraping tinsel popping up out of nowhere like an uninvited guest.

January or June, the high desert feels the same. Quiet as hell. The blank slate of snow or sand, teeming with unseen life. And Vegas, when it finally arrives, is the exact opposite: teeming with life on the surface, yet ultimately selfsame. The same air-conditioned sixty-four degrees indoors, January or June, the same cigarette-creased corners and widowed slot chairs and carpets hinting at possibilities stomped in long ago, the same high prices and poor odds, the tables forlorn, reminding you of someone else's long-spent hopes and riches.

Las Vegas, of course, is where you come to win money and leave memories. Or maybe it's the other way around. It's a city designed to disorient: the ritz of New York, the romance of Venice, the fertile valleys and pyramids of the Nile, the monuments of Greece. Vegas pulls you into a spectacle you can exit only much later, much lighter.

Vegas is also a paean to optimization. Its slot machines are engineered to keep you losing for as long as possible, with the jolt of a win just when you're ready to walk. The fantastical shows are decided by poll and priced by algorithm. The whole place is powered by electricity and water diverted long ago to turn desert into dollars. And those dollars, like everything else that happens in Vegas, tend to stay here.

For my part, I arrive in Las Vegas with some hope of understanding how we got here. Where did the technology of optimization come from, how did we wrangle the physical world into laser-thin efficiencies, and why did the lens of what's optimal come to dominate our worldview?

Most travelers don't get to Vegas via lonely highways like 93, and in many cases, they'd be hard-pressed to explain why they chose the place at all. Sure, there's the immediate reason: the wedding, divorce, birthday party, industry conference, or family vacation. Beyond that, it's unclear. Vegas is a thing one does, a place one goes.

Likewise, we're wont to ignore the whys and the how-we-got-heres of our optimized systems, instead accepting their algorithmic outputs as givens. The trains run on time, the phone calls go through, and the Google searches recommend results. Off we rush to work. We take out loans and buy in bulk and pay on credit. Yet a similar question nags, as it nags the Vegas visitor who steps outside the cool dark of the casino to take in the blinding desert plains: How on earth did we arrive here?

The previous chapter considered what we lost as we invited optimization in: slack, place, a sense of scale. Now, in Vegas, we pause to ask how these losses came about and what we may have gained instead. Optimization is the practice of pinning down an understanding of the world, of assigning boundaries and finding peaks within. This wrangling of the world into mathematical models required a set of conceptual shifts that settled into our mathematics over the past centuries.

The first shift was an increasingly atomized understanding of reality. The concept of indivisible units of matter and information—from subatomic particles to bits of code—turned the world into something to be divided, conquered, measured, and mined.

The second shift was toward greater abstraction. More and more abstract models—algorithms and analogies—allowed us to represent the world in increasingly complex ways. These models divorced representation from reality and placed control increasingly in the hands of those who could maneuver with such abstractions.

The third shift was toward increasing automation, applied first to the material world, then to the digital. Automation—whether a machine's stamping car parts or a software program's recommending news—decoupled output from human-scale input. It removed the checks and balances of small-scale, specific knowledge. It let us build more, more quickly.

These conceptual shifts inspired new mathematical techniques, from calculus and linear programming to machine learning and cloud computing. They gave rise to new infrastructure, from production lines and hospital-administration systems to server farms and search engines. By parceling and compiling, they sped everything up, and cemented optimization as a way of seeing. But while all this created immense growth, it kicked some important cans down the road.

VEERING OFF INTERSTATE 15 TOWARD THE NOW-REVIVING OLD downtown of Vegas, I arrive at a sleek-windowed building. It's an office designed not unlike a slot room, whizzing and brrrring and filled with bling, but it's far less smoky and more brightly lit. I'm not allowed in.

This is the headquarters of Zappos, the multibillion-dollar online shoe company, whose trajectory is peppered with examples of atomization, abstraction, and automation. Zappos was founded in 1999 in San Francisco. A giant fulfillment center outside Louisville, Kentucky, soon followed. The latter was located just a hundred miles from where, twenty years later, Amazon would break ground on its largest air distribution center in history. On the day it opened, the Zappos fulfillment center could hold a million and a half pieces of merchandise and process forty thousand per shift on "the largest carousel system in the United States," as the bubbly history on the company's website recounts.

I'm keen to tour the Vegas office, but it's the early days of the coronavirus pandemic, and I'm redirected to the online "experience," an assault of photos and corporate jargon. A friend later offers to introduce me directly to the CEO of Zappos, Tony Hsieh. Tony wrote a book called *Delivering Happiness*, which explains that by optimizing for positive experiences rather than dollars, you can actually create more dollars in the long run.

A few years after its founding, when Zappos officially moved its headquarters from San Francisco to Las Vegas, the majority of employees elected to go along. The rent was cheaper in Vegas, and although many were uprooted from the lives they'd built in the Bay Area, the move had a social upside, too. The workforce became more tightly knit as a result.

The move, by these accounts, was a win. Zappos was built to optimize for a set of values that it tracked meticulously as it grew.

While most companies measured their call centers' performance on average handle time, or the number of calls per person per day, Zappos looked at things through a "branding lens" rather than one of expense minimization. The company developed a one-hundred-point Happiness Experience Form to measure calls according to whether the agent made a "personal emotional connection," or PEC.

The goal, per Zappos, was to create as many "wow experiences" as possible. How does one create such experiences? It begins with simple, proprietary metrics to "Quantify and Reward Wow Moments."

On the Zappos website, Hsieh is exceedingly accessible, playing the part of a nerdy everyman. Here he is in a video outside his vintage Airstream trailer. Here he is wearing a fuzzy pink hat and conferring with the pet alpaca he's adopted. The everyman's pet. Tony talks about the future, which is ever-bright. The site seeps with early 2000s tech vernacular, the slang of an era in which techies believed they could speak to the whole world, and technology was all that was needed to save it.

Tony says of his alpaca, echoing the bright-eyed utopianism of Silicon Valley, "Pretty sure he's going to live forever."

There's ample room between the lines, of course, to read in a tone of calculated efficiency: move somewhere cheaper, save money on salaries and taxes and real estate. Disconnect employees from one place; move them somewhere new; the company becomes their social life.

Bit by bit, Zappos rolled out new optimizations, heartily promoted as fun. Hiring interviews were optimized to ensure the right cultural fit. In 2003, Zappos began to offer 365-day return periods to its customers. This, seen through the branding lens, was not just friendly service but also a means to maximize customer loyalty metrics and, ultimately, revenues.

Like many of its peers, Zappos is built on the language of industrial efficiency, first popularized by Frederick Winslow Taylor and later refined by proponents of Six Sigma and lean manufacturing. Its vision conjures that of Henry Ford, who strove for "ore to assembly," or end-to-end fabrication, at his Michigan plant. Ford wanted to transform raw material into finished automobiles, with no pause for downtime, no moment lost in warehousing or transportation to another facility. Zappos layers on some modern inventions, the language of agile software development, the buzzwords of HR. All optimized and wrapped up in a glitzy package of fun and teamwork and opportunities for growth.

When we see something that appears magical, integrated, seamless—a bunny out of a hat, shoes that arrive in forty-eight hours from a multimillion-item inventory, the razzle-dazzle acrobatics at a circus show on the Strip—our first impulse is often one of awe and marvel. Not far on its heels, however, is a second impulse: to deconstruct it, to understand.

Such was the impulse of one of the leading scientific lights of the past few centuries, and in many ways the opening bookend to optimization's story.

Newton is known for many things, from the mathematics of calculus to his apple-on-the-head discovery of gravity as a universal constant. He did important work in optics, showing that white light is a composition of spectral colors, and in physics, detailing the laws of motion. Perhaps more than anything, Newton is remembered as one of the key players in ushering in a new age of rationality and a new way of doing science.

Indeed, Newton's ideas helped form a foundation for a more experimental style of research. But much of this foundation would be built out only later in his life. The retrospective fairy tale forgets that Newton stood not so much at the beginning of one era as at the end of another and that especially in his early years he was primarily consolidating an older way of understanding.

Newton was fascinated by alchemy and strove to find that "material soule of all matter," the "secret fire" of nature. John Maynard Keynes, in his 1946 address to the Royal Society (delivered posthumously by his brother) wrote: "Newton was not the first of the age of reason. He was the last of the magicians, the last of the Babylonians and Sumerians, the last great mind which looked out on the visible and intellectual world with the same eyes as those who began to build our intellectual inheritance rather less than 10,000 years ago."

Many biographies exist on Newton, and his character is shaped as much by the times in which historians were writing as by the actual man. Lately, he's been reenvisioned as a boy genius, with flowing locks, lounging in the orchard when a falling apple flashes the gravitational constant into his mind. Or as biographer William Newman puts it, a "Caltech geek," a mild autist with savant-like qualities. Others have called him nuts; biographers writing in the 1960s and 1970s, taken with psychological analysis, adopted Keynes's view that Newton was, more than anything, "extremely neurotic."

What all agree upon, however, is his wild and wide-ranging intellect. Keynes goes on to cite examples in which Newton long knew something based on intuition, such as his notion that the pull of a spheric mass could be approximated by a point concentrated at is center, then hurried at the last minute to provide the formal proof needed. "His experiments were always, I suspect, a means, not of discovery, but always of verifying what he knew already," Keynes wrote.

Obsessed with the secret recipes of alchemy, young Newton's quest was to turn experimental findings into formulas. As a result of this consolidating look backward, and perhaps almost by accident, his work marked the beginning of a new era of science. Isaac Newton didn't invent optimization, but he serves as a useful milestone for approximately when this way of thinking started to take hold.

Although Newton's interests hearkened back to earlier, unresolved mysteries, it was by dividing the world into smaller parts—parts that would eventually form the basis for his calculus—that these mysteries could begin to make sense. The ancient Greek Leucippus and his student Democritus are credited with one of the first conceptions of a material world made up of selfsame units. These units maintained important properties of the whole. An atom of iron was thus heavier than an atom of air. A bird or a stone comprises smaller pieces of its essence. Aristotle later described the world as made of four elements: earth, air, fire, and water. His concept of "minima naturalia" described the smallest parts into which a substance could be divided and still retain its essence.

As later philosophers studied the topic, some questions arose. How can something be, in theory, divisible an infinite number of times and yet, in practice, also have a definable smallest unit? Medieval thinkers such as Avicenna considered and ultimately resolved the paradox by proposing a distinction between physical and conceptual divisibility. That is, we can represent the physical world as infinitely divisible, even though there exists a set of smallest units. This distinction anticipated not just a more modern conception of the atom but a new approach to physics.

After his excursions into alchemy subsided, the second half of Newton's life was one of relative staidness. He retreated into university business and private dinners, paperwork and wine. Although he was passed over for a number of academic positions, his income was secure. He continued to invent but also grew fat, beleaguered, and per some reports, "gaga."

Yet it is for his staid later work that Newton is best remembered. The physical world is not unpredictable, says his way of seeing, but can be harnessed through careful measurement and calculation. You can dice the curved path of an object into tiny slices and sum them back together to know how far it's traveled.

And what happened in the sciences over the centuries following Newton was nothing short of transformative. By the middle of the 1800s, physicists were shoving electrical current through vacuum tubes and shining light across double slits to figure out if it was made of waves or particles. By the early 1900s, they'd developed ways to measure and weigh and started to name these tiny units. Atoms were not, as Democritus had posited, the smallest of the small. There were smaller units yet. Up and charm quarks. Photons and Z bosons. In 2012, the Large Haldron Collider, at an operating budget of over a billion dollars per year, would provide a glimpse of the Higgs boson, the most elusive subatomic particle to date.

In concert with these advances in physics were advances in the specialization of machinery and of society itself. In the famous description of the pin factory that opens *The Wealth of Nations*, Adam Smith describes the atomization that allows each to focus on one part: "One man draws out the wire, another straights it, a third cuts it . . . In this manner, divided into about eighteen distinct operations, which, in some manufactories, are all performed by distinct hands."

Adam Smith's vignette recalls the often overlapping precepts of optimization and market capitalism, and it's worth pausing to draw comparisons. Smith's pin factory, Zappos, and Ford Motors are for-profit companies, with the goal of maximizing dollars; in this way, they are optimizing. An optimization, however, can undergird much more than revenues. It could be profits, or it may just as well be speed, or in the case of Zappos, some quantified notion of emotional connection. It might be what economists call productivity, that sometimes-elusive division of total output by total cost. It could be the "incremental cost-effectiveness ratio," which in the United Kingdom is used to evaluate whether a new health measure should be implemented.

In this sense, an optimization is agnostic to its context. Whether we want productivity or emotional connection, sweet lemonade or sour, the mathematics is the same. Capitalism, on the other hand, is not just an operating mechanism but a political philosophy that favors profitability as a measure of success. The market—that is, the aggregated preferences of customers over time—decides which businesses and which ideas survive.

More than this, market capitalism and optimization differ in how they ultimately land on outcomes. The drive of optimization is to hem in, to bound and control, in order to find the best. It begins with an answer, an objective function to fit, and offers a process to maximize therein. Markets, on the other hand, don't so much begin with an answer as set things in motion with markets and regulations for transacting. Capitalism is bottom-up, while optimization is top-down. Market capitalism says, The purpose of open exchange is to allow for the best to emerge; we can't dictate the end, only the rules of engagement. Optimization says, Given we've already fixed our idea of best, how do we work within our constraints to get as close to that as possible?

In the early nineteenth century, the development of machinery brought about stamping plants that could make thousands of the same part, over and over. Tools such as lathes, planers, jigs, and fixtures could be reassembled into different configurations. One such plant, in Buffalo, would open in 1950 and eventually grow to cover eighty-eight acres and ship out one hundred railcars of auto parts each day.

As chronicled in narratives such as Upton Sinclair's *The Jungle*, the meatpacking plants of Chicago helped bring this mechanized system to life. Workers sat at stations, where a pulley system delivered the meat, completing one task over and over. Sinclair describes the heartbreaking inhumanity of mechanization, to animal and human alike.

Henry Ford would later cite the meatpacking industry as an inspiration for his own factories. William Klann, a machinist who rose to be Ford's head of production, introduced the assembly line after visiting the Swift & Company slaughterhouse in Chicago. Ford's set of principles—interchangeable parts, continuous flow, division of labor, and reduction of wasted effort—were modeled on what Klann saw there. In *My Forty Years with Ford*, early Ford manager Charles

Sorenson describes how this focus reduced the time to complete an automobile about eightfold. It was so transformative that the bottleneck in the process became the auto paint, which just wouldn't dry fast enough.

WHEN ZAPPOS MOVED ITS OFFICE FROM THE SUBURBS INTO THE FORMER city hall of Las Vegas, in 2013, it felt like a turning point. Four years earlier, in 2009, the company had been quietly purchased by Amazon. The acquisition was referred to internally as a partnership, with a riff on the nursery rhyme: "Zappos and Amazon sitting in a tree." The shoe seller wanted to maintain its separate brand, and Amazon agreed. That brand was profitable.

Its profitability, moreover, could be measured, linked, and atomized. Every customer-service call was rated on its human-centric metrics. Each shoe style was reduced to the same set of features— brand, size, color, height, material—to allow for better sorting and search on the site. New updates to the site were A/B tested to ensure the optimal design. Every line of code came down to a set of 0s and 1s. The call centers buzzed away with the fast-careful tempo of a machine. The human touch that set Zappos apart from its parent company was in fact quite carefully orchestrated.

Las Vegas itself is the result of smallest units, or atomic building blocks. These units are combined and recombined, forming new outcomes each time. They're the cards and the chips, the coins in the slot machines, the specialization of a labor force into bartenders and dealers and circus acrobats. Atomization is what enables the computers to run, identical and in parallel, deployed to calculate odds and dole out prizes.

Indeed, it's the same atomization that underlies much of computation in general, held together by the glue of abstract expressions and formulae. When we start to see the world in terms of selfsame units, of 0s and 1s, of call-center metrics, of electrons that can be peeled away and pasted onto new substrates, of a genetic code that can be spliced at will, we start to see a world that's ours to build. Across various disciplines, this atomization of our material reality created a division between part and whole, and cast to the side those things that weren't readily divisible. The result was a reductionist and blinders-on science, economy, and culture. This chopped-up view is a far cry from Democritus, whose project in seeing the part was a distilled expression of the whole.

HERE IN A DINGY VEGAS HOTEL BALLROOM, THE WORLD SERIES OF Poker (WSOP) is in full swing.

This is no glamorous shot from James Bond. There's no sweeping footage of Montenegro casinos and highways and sports cars; there are no glamorous women dropping poison powder into dry martinis. In fact, there are few women at all.

Instead, there's a mercantile feel to the whole affair. The hallway lights are bright, too bright, the carpet's wearing thin, a booth hawks sunglasses with dark lenses to hide expressions. Outside, Uber drivers in Priuses whisk downtrodden players away, as the reality of losses unfurls across the hot concrete. It's early July, and triple-digit temperatures are in the forecast. Here and there a superstar has nabbed a small conference room to do a live-cast. A handful are dressed outlandishly. Most wear black or neutral fleece sweatshirts, with their sponsor's logo patched on. The gender ratio of dealers is far more balanced than the players'; in the restroom, I wash my hands between two female dealers reapplying lipstick.

The year is 2018, two years before my pandemic swing through town. At the tables, eyes are shaded by glasses or visors, leaving only the players' mouths to cast the occasional outburst or spittle across the table. It's hard to get a glimpse of the action, but the bigger final rounds are recorded and streamed.

Poker is different from the slot machines. There are a set of probabilities for each hand, out in the open rather than hidden inside a whirling box that spits out coins. If the result of atomization was a flattening, a commodification that allowed pairs of shoes or the probability of aces to be treated all the same, in games such as poker, we see how abstraction was then built on top.

At events of the WSOP's caliber, pretty much all players have the odds down pat. What top players layer on further are models of other players. They read people's faces, study up on their psychologies and previous play. A young woman at the table, let's call her Nell, is one of the series's best. Her frame is slight but fierce; her insults haughty and biting. Today, she's set out to win a bracelet in the $5,000 no-limit Hold'em.

I meet Nell through a college roommate who went on to become one of poker's top players, before retiring in the late 2010s. In college, my friend would often disappear into basement tournaments or online games and, later, to the Tribal casinos nearby. Even after stepping back from the game, her world is shaped by probabilities. She's started to see her nonpoker relationships in game-theoretic terms. If he does this, she'll do that, unless he first does this.

To my friend, the best was always of utmost importance. Although she's since relaxed her ways, in a telling moment, she once revealed a teenage obsession with the computer game Minesweeper. In high school, she was ranked number two in the world at the game. This depressed her to no end. "I just don't understand the point of playing if you're not going to be The Best," she'd say.

Like any child of the 1980s with access to a PC, I, too, had dabbled in Minesweeper. That the point of playing could be anything other than a mildly challenging, often-mindless distraction struck me as a wild idea. But the more I sat with my friend's comment, the less I could come up with any counter. The best was something clear, concrete. Anything less than that lacked definition, an unmoored dinghy in a muddled sea. To be second may as well have been to be a million-and-second. If not the best, what else was there?

IT'S PERHAPS UNSURPRISING THAT THE IDEA OF QUANTIFYING THE best has its roots in Vegas. Ed Thorp describes the first time he met Claude Shannon, by then a world-renowned mathematician and fixture at MIT. Thorp was a young instructor and revered the slightly older Shannon, who was already famous for his work on switching circuits and the logic at the root of digital computers. After several inquiries, the latter reluctantly agreed to take a meeting, if they kept it short. As recounted in Thorp's autobiography, *A Man for All Mar-*

kets, the younger man scurried in at the appointed hour, a bundle of nerves. Against expectations, the two hit it off, and the allotted ten minutes turned into several hours.

Now fast friends with a common interest in odds, in the fall of 1960, the pair set off to design a system to beat roulette. They bought a $1,500 regulation roulette wheel and spent hours in Shannon's basement, which Thorp describes as "a gadgeteer's paradise with perhaps $100,000 worth of electronic, electrical and mechanical items." Their final invention involved a small computer around the waist and a switch to tap with the toes, as the ball circled the roulette wheel. The computer clocked and calculated the ball's movement to generate a prediction, which was communicated via an eight-note musical scale to an earpiece, relaying approximately where the ball would land.

After several months of perfecting their invention, Thorp and Shannon and their wives met in adjoining hotel rooms in Las Vegas and wired themselves up with the counting mechanism. Shannon joked to Thorp, with the switches click-clacking under their shirts, "What makes you tick?"

This was, per several biographers, pure Shannon: a man in love with games. He'd already made his fame on circuits and probabilities, on the theory underlying chance, and now games were going to comprise his semiretirement. Although the roulette invention was cumbersome and thus short-lived, his attention continued along similar avenues. Per author James Gleick, "The open secret at MIT was that one of the greatest minds of the twentieth century had all but stopped doing research—to play with toys."

Shannon, who grew up in Michigan, studied mathematics and electrical engineering, first at MIT, later at Princeton's Institute for

Advanced Studies, where he crossed paths with the likes of Albert Einstein and Kurt Gödel, and then at Bell Labs, where he worked on cryptography directed at wartime efforts. But Shannon's big innovation, and the reason he could kick back and play, was enclosed in a shortish paper titled "A Mathematical Theory of Communication." The paper stems from a simple question: What is the best way to encode a message?

Prior to this work, the question of communicating information was treated as a deterministic one. Take a phone call with background noise. The sender has a message in mind, and the receiver has to do his best to reconstruct it. For example, if I were to call you from a loud bar, and say something that sounds like "Meet Mim at the hostel," you'd have to figure out where to go. And whom to expect to meet when you get there.

With the noise of the bar, however, the signal is unclear. Mim might refer to "him" or to "me" or to someone named Mim. Likewise, "hostel" could be a hostel or a hotel or the hospital.

Shannon reframed the problem by identifying a key element on the receiving end: that of uncertainty. You may not be sure whom you're meeting or where, but there's less uncertainty in the middle part of the message, the fact that I'm asking you to meet someone. In fact, the total uncertainty of a message can be quantified probabilistically by a measure Shannon calls information entropy. The higher the entropy, the more uncertain you are about my meaning.

This principle applies to all kinds of communication. A phrase like *hahahahaha* is relatively easy to encode and therefore low entropy: it's five ha's in a row. If a singer repeats a chorus three times, or sings the letters of the alphabet in order, it's also pretty straightforward to describe the information content, or the difficulty of conveying the message to someone else. But a string of randomly generated nonsense words, or the babble of a three-year-old with an imaginary friend, would be somewhat more difficult to encode. The entropy of a message is a matter of its complexity rather than its length. I could chant *hahahahaha* for six weeks on end, and it'd still be less complicated than a three-year-old speaking nonsense for just six minutes.

This was Shannon's breakthrough: to describe information as ultimately atomizable, and to define informational complexity with a measure stolen from the physical world: entropy. In so doing, he layered an important abstraction on what had been seen, before, as a deterministic act of recovering a message. He also laid the foundation for much of computing, compression, and a new field called information theory.

LIKE ANY OTHER VISITOR, I, TOO, HAD SURFACE ANSWERS FOR WHAT I was doing in Vegas. I was there to watch poker, to drive by the Zappos office and down the Strip, to feel the quiet of the early pandemic against the backdrop of past boister. I'd read Shannon's paper at a young age, and my work in graduate school was related to information theory, using a concept called Kolmogorov complexity to show that the data traces describing human behavior do not so easily conform to tidy rules. People may be predictable at the large scale, but when you zoom in, we retain some small upper hand against the models.

A model is an abstraction, a consensus shortcut we use to communicate. A painting of a wolf is an abstraction of the live animal. We can use it to tell a story, to invoke the spirit of the animal, without having to capture an eighty-pound beast and lug it home to bedtime. The word *grapefruit* is a shortcut to describe an object, a color, a texture, a taste. The number 7 can represent anything from deadly sins to notes on a scale, that can be summed or multiplied per the abstract functions of arithmetic.

Abstractions allow us to connect cause with effect: we might say that when there are dark clouds in the sky, it's likely to rain. They also let us suggest new patterns of pass along a point of view.

Claude Shannon took atomized units of information and added a model to describe how to quantify the uncertainty in a string of such bits. As computers grew more sophisticated, programming languages and interfaces and remote servers processed far-off computations via the abstraction we call the cloud. Drag-and-drop interfaces put thousands of lines of code behind the scenes, so that a noncoder could make a website.

If we compare these modern abstractions—such as a tablet computer or the financial models of Wall Street—to the representations and stories of yore, a few things stand out. Our newest abstractions encourage not just simple description but also prediction. And they involve many more layers of complexity. As Matthew Crawford writes about the modern digital world in his book *Shopcraft as Soulcraft*, "In becoming less obtrusive, our devices also become more complicated." So it is with our models.

In other words, a modern abstraction often stands in place of the pattern it describes. A bus schedule, a demographic model, a climate forecast, a credit-default swap—these all take observations and string them into a pattern. The more we rely on the pattern, the farther we stray from observing.

IN 1831, A YOUNG FRENCHMAN CAME TO AMERICA TO STUDY THE budding nation's prison system. He was on assignment from the king and was eager to explore Alexis de Tocqueville was just twenty-eight years old, and his mission quickly expanded in scope. He went on to travel for nine months, talking to people of all sorts and capturing the paradoxical and parochial parts of a country's early history.

Tocqueville's resulting journals comprise a tome, almost biblical in range. The writings sprawl out like a Vegas casino, avenues of anecdote and sweeping rumination. And as with the Bible and Las Vegas, certain themes repeat. One of Tocqueville's favorite topics of inquiry is what makes Americans so peculiar.

A driving element of our peculiarity, he claims, is found in our fondness for abstraction. Indeed, we're better at it as a result of residing in a democracy, having a mistrust of aristocracy, and seeing others around us as created equal. "The man dwelling in a democracy . . . is aware of beings about him who are virtually similar; he cannot, therefore, think of any part of the human species without his thought expanding and widening to embrace the whole."

A modern-day Tocqueville might be astounded to find, in the middle of the desert, the Eiffel Tower and fountains of the Bellagio. But he might be less surprised to see Nell at the poker table, peeking at the corners of her cards and pondering an abstraction.

Poker is a game made possible by atomization of chips, cards, rounds, and hands. And by abstractions: the rules of the game, the combinations and probabilities, the mental models of other players, and, increasingly, backers and sponsors and money and debt.

The complexity of these layered abstractions is evidenced by the time it took for a computer to beat humans at poker, compared to chess. When Deep Blue bested Garry Kasparov in 1997, it was a triumph for computational power more than anything else. The computer had only to brute-force a search through all possible future moves, then take the move that maximized its chances of success.

Not so with poker. As my friend and Nell well know, the probabilities themselves are merely a first step. Good players form mental models of what others will do to gain an edge, studying opponents' strategies—and especially their weaknesses—in past tournaments. They also make sure to remain unpredictable themselves, by playing it straight sometimes if they have a reputation for bluffing and vice versa. It took two decades after Deep Blue for a computer to beat humans at poker, and even so, it can't win consistently. It can do so only in expectation: with infinite games, the computer will come out ahead.

The key innovation in the 2008 algorithm that could beat the pros at heads-up Texas Hold'em came out of a Canadian university team called, appropriately, the Computer Poker Research Group. The algorithm used what's called a regret minimization strategy, calculating probabilities for every possible hand, then making a betting decision to minimize the deviation from an optimal strategy, given the resultant hands and bets of other players. This, you might guess, is no small feat.

Over the course of two months, the algorithm played against itself a billion billion times, on four thousand computers at the rate of six billion hands per second, updating its knowledge with each hand. The team's lead researcher, Michael Bowling, points out that this represents more games of poker than all of humanity has ever played.

Some years after this, in 2015, my friend went into retirement. She said the game had changed; too many people now know the probabilities and play by odds alone. The community has also changed, from small and intimate to large and impersonal. Those of the old guard left now see tournament play as equating to an hourly wage. Poker players, more than any professionals I've met, tend to think in expectation—that is, they map future possibilities and their associated odds and make decisions based on what they'd gain from an outcome, multiplied by the chance of it happening at all.

My friend left because the game was "solved." It had become optimized and, therefore, boring.

Still, she hangs around the tournaments and revels in the excitement of live play. At the 2018 WSOP, until the last minute, it looks like Nell will win the round and advance. Then, she's dealt a really bad hand. She bluffs. A player across the table calls. She's out.

That same year, the marriage of Zappos and Las Vegas was consummated in a new way. A carefully crafted PR blurb tells it like this: "Zappos takes a first step in disrupting the entertainment industry with world-class customer service. In collaboration with Caesars Entertainment, Planet Hollywood's Axis Theatre on the Las Vegas Strip was rebranded as the Zappos Theater."

The thousands of annual visitors to the Axis Theatre who come to see Shania Twain or Jimmy Kimmel hardly notice its new brand identity, just as they couldn't point you to the old city hall, or tell you that their shoes were bought from an Amazon subsidiary. Like the optimizations that enabled it, it was a quick and largely silent change.

THE ZAPPOS THEATER AND THE WORLD SERIES OF POKER MAY BE showy examples of a certain flattening, the combination of atomization and abstraction into something at once big and largely predictable, but the roots of this flattening run much deeper. They trace back, perhaps, to a Frenchman's wager. Not long before Newton was making his alchemical inquiries around 1658, Blaise Pascal made a bet. Unlike a poker player holding her cards close to the chest, Pascal made his demonstratively and in writing.

Pascal, a mathematician and philosopher, lays out the following thought experiment. Either God exists, or God does not exist. Moreover, God's existence is something reason can neither prove nor disprove. Therefore, faith in God can be considered as a wager of sorts. You can choose to believe or not.

Pascal examines the expected outcome of each choice. If you choose to believe and God does not exist, you lose very little: maybe some time. However, if God does exist, you stand to gain an eternity of bliss in Heaven (presuming entrance to the afterlife is predicated on positive belief). On the other hand, if you choose not to believe and God turns out to exist, you face a near-infinite loss in the afterlife.

It was a thought experiment, and a pragmatic one at that. Pascal's phrasing suggests an early type of game-theoretic thinking. The philosopher of science Ian Hacking, in his impressive overview of the history of probability, describes the wager as "the first well-understood contribution to decision theory."

Ian Hacking goes on to call Pascal the first statistician, laying the foundation for modern probability theory and ultimately optimization. What Hacking means by this is that Pascal was among the first to codify a systematic way of looking at possible outcomes. Pascal's wager was an example of what atomization and abstraction allowed: a forward-looking prediction of odds. By aggregating particulars, we could make a guess about the future.

Over the next centuries, mathematics was built around both Pascal's probabilities and the calculus Newton and Leibniz had invented. Eighteenth-century polymath Leonhard Euler, who developed the mathematics for finding the high or low point of a curve, wrote that "nothing at all takes place in the universe in which some rule of maximum or minimum does not appear."

As we became more disposed to predict, we lost something, too. The game became, as my friend put it, "boring." It was flattened, commodified. The mystery and the fun were gone. The same flattening made Vegas, Eiffel Tower notwithstanding, into a facsimile of so many other US cities: arrive and rent a car from one of the same conglomerates, buy a drink at Starbucks, stay in one of the hotels owned by the same few chains.

Pascal's was a trade with the future in more ways than one. In the obvious sense, he was toying with his future odds of salvation. But in a deeper sense, by introducing his game of what-ifs, Pascal cemented a present calculation about some future odds. He traded an increased ability to reason and make wagers in the present for a loss in future flexibility. Any big bet is like this. You accelerate forward your expected gains or losses, pinning them on a moment close at hand.

Optimization is also like this. We've seen before how optimizing involves trade-offs, many of them hidden. And we've seen how it cements a certain way of seeing. Let's say, for example, you have the job of parceling up some portion of the Nevada desert for its first homesteaders. You have some choices to make now, and while they seem arbitrary and flexible in the present (Should I create ten-acre plots? forty-acre? one-hundred-acre?), once decided, they become harder to reverse.

Let's say you decide on forty-acre plots. You hire a surveyor to walk the land, a cartographer to draw it all up. Now the map is made. You cannot simply say, as you could have when you began, I want ten-acre plots instead. Now that the map is fixed, you take it to the printer's shop. A hundred copies are made, a thousand. It makes its way into people's hands. Young men, and old, take the train out from cities far in the east to choose their forty-acre plot. They dig wells, build cabins, graze sheep. The lines of the map are fixed in their minds. There's really no going back now.

FOR SOME, IT IS THE VERY ELEMENT OF SURPRISE, THOSE PARTS OF Vegas that are still human, still imbued with human emotion and thus gameable, that keeps it fun. For others, it's the flattening, the emotionless algorithms behind the slot machines, that keeps Vegas safe and keeps them coming back. You are bound to lose, in expectation, over time, but at least there are no surprises. Harrah's CEO Gary Loveman praises his casinos' perfect analytics for tracking odds. He "sees his customers as a set of probabilities wrapped in human flesh." Everything is automated.

If it were a Hollywood blockbuster, the highlights reel of automation would be composed of scenes from the machine age, with its churning machines and loud clacking noises and metal monoliths, the idea that we can repeat anything, to lighten the human load. The first computer was envisioned as a mechanical engine, and it quickly got smaller and smaller. We made millions of them; we made them so small they'd fit in a pocket and could be turned on by a fingerprint.

Mathematician Merrill Meeks Flood took a centuries-old question, known as the traveling salesman problem, of how to find the most efficient route between a set of points on a map, as arises, for example, when a salesman goes door-to-door or a bus needs to make a set of stops. Using an automated approach, Flood worked out a new way of routing school buses through neighborhoods, developing the algorithms that would come to underlie modern scheduling software.

Automation is what allowed Henry Ford's River Rouge project to take off. The factory's long, quiet avenues are not unlike Vegas's boulevards, its hidden factories full of motion like the bustling casinos shaded from the desert sun. Ford's vision wasn't simply one of mechanized labor. It was also a vision to control the entire supply chain, to harness both money and time.

Zappos's customer-service phone calls, factory lines, poker earnings—it's very easy to lose track of scale. Suddenly you don't know every employee's name, suddenly you go from winning to losing fast, suddenly the supply chain you thought you controlled is chaotic. Or maybe the water will dry up, the lights will flicker, and the power will go out. Not for long, but just long enough to remind you you're in the desert.

Part of Las Vegas's beauty results from its improbability. Here is this splendor, this electric spectacle of light and power, which would cease to exist if you merely flicked off the switch that keeps the water churning through to power the lights. Not slowly, not gradually, but all of a sudden: the casinos would go silent, the lit-up sky turn black, the swimming pools evaporate.

It could just as easily not have existed. The city was born as a stopping-off point for Spanish traders en route to Los Angeles from New Mexico. There were artesian springs in the valley; the surrounding plains were green. It took a great deal of enterprising, down the line, to wrest more water into the valley, and still the city doesn't yet have the water it needs.

Las Vegas, in other words, is one giant can kicked down the road. Just southeast of the city is a sparkling body of water, with many beneficiaries and many masters. Since it was formed by the creation of the Hoover Dam in 1935, Lake Mead has been the largest reservoir in the United States, providing water to some twenty million people and vast acres of farmland.

On a warm day, and even on a cool one, small streams of families make their way down to the water, carrying floaties and coolers. During pandemic times, groups keep their distance, toddling children called back when they wander too close to another bubble. Music from stereos drifts defiantly between encampments and occasionally a small dog will escape, to sniff another or the contents of a cooler.

The Colorado River was once called the American Nile, for its

dramatic flooding of the surrounding valleys, alternating fertile wet-
ness with parched dry. The Hoover Dam put an end to the floods
with seven hundred vertical feet of concrete, expected to hold up for
millennia.

Building the dam diverted massive quantities of water, turned
valleys into lakes, and forced out an entire small town in order to
make space. In conjunction with the reservoir came the Law of the
River, also known as the Colorado River Compact, which defined
allocations of water among the western states.

All this ground shifted, water diverted, to alleviate certain griev-
ances and set others in stone. Prior to the compact, the states of the
Upper Basin—Colorado, New Mexico, Utah, and Wyoming—feared
California, its population exploding, would deplete the rivers that
flowed through their boundaries, and upon which their residents
relied. The Lower Basin states—Arizona, California, and Nevada—
were likewise worried they wouldn't have enough water to reliably
develop their agriculture or cities.

The compact was drafted and signed in Santa Fe in 1922, presided
over by Herbert Hoover, then the secretary of commerce. Careful
calculations were done of river flows, and an agreement took shape.
The Upper Basin states would not deplete the flow of the river to less
than 7.5 million acre-feet (about the amount required to cover all of
Los Angeles in three feet of water) in any ten-year period. In surplus
years, the Lower Basin states would receive an additional 1.1 million
acre-feet.

Allocations were based on observations of rainfall in the years
prior to the treaty's signing. Everyone would get their fair share. It
was a great idea, in theory.

In practice, there were a few complications. When the compact was signed, no one figured in the losses of water due to evaporation, which are substantial in a region known for its sunshine and dry climate. Some 1.2 million acre-feet thus disappeared annually from the ledgers into thin air. Subtract another chunk gulped up by plants in the surrounding riverbeds. And then in 1944, a new treaty allotted 1.5 million acre-feet to Mexico, without specifying where it would come from.

In addition to the Law of the River, there were compacts developed between individual states, and within-state rights of seniority. Some places have more than a dozen claims on a drop of water passing through. "In the West it is said, water flows uphill toward money," writes Marc Reiser in *Cadillac Desert*, his history of the western United States.

In 2020, Lake Mead's water level stood just above 1,075 feet. If the reservoir drops below this level, Arizona loses one fifth of its allocation, with farmers and ranchers bearing much of the brunt. By the late summer of 2022, the Department of the Interior announced a series of cuts, as well as a scheme to pay farmers and cities in Arizona, California, and Nevada to voluntarily reduce water usage. The longer users agree to cut their water, the more they will get. A one-year contract pays $330 per acre-foot, while a three-year contract pays $400.

If the lake drops below 895 feet, it enters a deathly spiral, or "dead pool," with water below the level at which it can exit. The lake is shaped like a funnel, so the lower the water level gets, the faster it goes down. Below a certain level, it's not just water contracts that won't be met. It's also power.

For you and me, for most creatures on the planet, water and energy are not abstractions. Without clean water, our bodies shrivel, our gardens die. Without heat, we freeze. For the vast majority of Americans, whose primary source of power is the electrical grid, without power they can't heat or cool their homes.

For the states' representatives signing the original compact, water was a real thing. But it could also be abstracted away, such that it was no longer two men fighting over a single pail of water. It was no longer a local farmer objecting to a dam that would flood his land, against a downstream farmer who wanted his fields irrigated. The signers thought about the future, of course, but it's hard to legislate the unforeseen. The compact was signed when rainfall was higher than average. It was numbers on a page, a system that could be wrangled, whose agreements could be backed not just by one town's armed men against its neighbor's but also by the full abstract force of the law. It could even be traded for other numbers on other pieces of paper.

Lake Mead hasn't seen a surplus in nearly twenty years. In fact, the majority of the past two decades have been drought.

AS I WALK THE SHOP-LINED CANALS OF THE VENETIAN, A TWIN FEELing emerges: of possibility and emptiness. Vegas offers promises of bounty, yet there's also something that feels hollow, in conflict with itself.

In parallel with increasing abstraction, the machine age gave way to an algorithmic one. New software allowed giant optimizations across systems with which an operator might not be intimately familiar: the electrical grid, for example. Cloud services reduced the price of computation by orders of magnitude, transforming what was once accomplished in mechanical (and also very human) timescales to lightening speeds. Just a hundred years ago, we could barely measure daily weather in most places; by 2020, anyone with a laptop and a few bucks could run a simulation of future climate patterns across the entire globe.

But something else happened when we went from automation in the physical world to automating computation. When we mechanize not just a physical process (the sawing of a board, the pounding of a hammer) but also computation itself, things change. When robots build cars, the process is deterministic and linear. Ten robots can build one car in twelve hours, let's say. And we know what car they're going to build.

But when we scale up algorithms, the output decouples. A news recommendation algorithm can generate ten recommendations per hour, or it can generate ten billion. Ten billion may require only a fraction more computing power.

It also, often, becomes nondeterministic. On a factory floor, one procedure, such as stamping out a part, maps to the same output every time. When the first algorithms were described, they were fairly simple: input X to get Y. By contrast, a present-day algorithm to recommend news articles does so probabilistically. A browser might surface one set of results today and another tomorrow, with no major change in the algorithm. No human could manually track the optimizations across the scores of features that are occurring to serve such articles and ads.

Peter Drucker predicts, in his 1993 book, *Post-capitalist Society*, that the Information Age will have humans telling machines what to do and setting the pace, in lieu of the machine pace that characterized the industrial age. Part of this has come true. Despite salacious stories of artificial intelligence (AI) takeovers, humans still largely tell machines what to do. But they often do so in a way that's disconnected from pace, place, and effect. This disconnect is epitomized, for example, in chilling drone strikes on the other side of the globe, controlled like video games from military chambers within our borders. We're disconnected not just from the simple pleasures of life but also from its brutality.

It's worth remembering, in all this, that there are many things made possible by optimization. New efficiencies in tools and ways of living and making do from the earth's resources have created all manner of marvels. Without optimization, we wouldn't have interstate transit, we wouldn't have Los Angeles, we wouldn't have modern life spans and the variety of food available worldwide, we wouldn't have much of modern medicine and dentistry. We wouldn't be able to speak with someone across the globe through a computer screen. Whether you believe these developments have been for good or for ill, they wouldn't have been possible without taking advantage of efficiencies in a way that nature does not, through human engineering.

And whether or not you appreciate optimization's artifacts, it's equally true that with all we've created, we've lost a great deal.

These losses run deep. We've lost track of where our money goes. We're not sure how the glimmering lights of Vegas stay on. We've forgotten about the diversion of water a century ago, which allows the servers to run, the call centers to buzz, the slot machines to turn, the airplanes to fly.

As our daily routine churns ever faster, it grows threadbare, pulled on every side by machinery, loans, advertisements, schedules, addictions, obligations. We feel numb and powerless, as if having lost the capacity for invention, imagination, and connection. These are the very things that allow us to outdo the machines, and yet in our threadbare numbness we slap on more patches to keep the machines buzzing just a little longer.

The quietest month in Las Vegas is December. Nell has gone home. The big tournaments and conferences have wound down. Out trickle the pros, leaving an ocean floor of desperate gamblers, of stray families and bachelors, grandmothers in pairs or alone. The slot machines whiz and the craps and blackjack tables ebb and flow.

I traveled to Vegas to understand where optimization came from. I wanted it to all add up, to stuff it into a story that made sense. Instead, I'm left with a thousand bits and pieces, confetti floating on the fractured air. As I leave Las Vegas on the interstate at dawn, light separates from dark and desert mirages replace the hyperreal simulacra of the Strip.

3

HIGH DESERT CHURCH

JUST PAST THE PARCHED TOWN OF VICTORVILLE, CALIFORNIA, SITS A quiet shrine to efficiency.

Fenced in, in the middle of nowhere, a dozen 777 jets are parked in perfect formation. These are part of the Air New Zealand fleet, which once made regular trips between North America and Auckland. When the coronavirus broke out in early 2020, borders were closed, and international air travel came to a screeching halt.

The 777 is one of the largest aircraft in history, and the planes possess a hulking, quiet elegance. One 777 can carry more than three hundred people on two stories, whizzing around the globe at some two thousand miles per hour. A typical flight has three to four pilots, who trade between rest and shifts in the cockpit. These giant planes are grounded until further notice.

Down the road is a larger airplane boneyard, a hint of what might become of the 777s. Here, not one hundred miles from downtown Los Angeles, airplanes are left to bake in the sun, with no plans to be fixed or parted out. Like much of our optimized world, they're ruins hiding in plain sight.

These ruins remind us of not just the conquest but also the costs of optimization. In the air, the planes allowed us to forget place, slack, and a sense of human scale. As ruins, they can be forgotten altogether, except perhaps as a curiosity, a detour on a road trip to Los Angeles.

The technology that forged the planes, and the culture that stuck them in the desert, are two sides of a coin. The previous chapter described how optimization came to be via three conceptual shifts— atomization, abstraction, and automation. This one considers how a set of cultural shifts has cemented optimization as the dominant mindset of the modern West.

One of these shifts is toward a more individualized, rather than institutional, access to knowing. We like to know for ourselves, or as Alexis de Tocqueville put it: "[Americans assume] control independently for their own judgment ... They are used to relying upon their own evidence." A founding myth of America is not just a monetary and military independence from the old world but also an independence of thought. The idea of self-sufficiency has long pervaded our frontier mythology.

The second shift is toward the idea that it is within humanity's power, and indeed its duty, to create a more perfect world on earth. As we'll see in this chapter, a set of nineteenth-century British thinkers helped to codify how this could be done programmatically and through individual action. This helped cement optimization as a lens of master planning, connecting part to whole in an organized way. What had once been in the province of private virtue was now a public means, a tool. Against the backdrop of an industrializing society, this tool became an increasingly agreed-upon standard for governance and social norms.

The third shift is toward a codification of these two ideas: individual action *in service of* a better world. While many of America's big inventions, from the interstate highway system to the supercomputer, may well have occurred without this mindset, they would not loom so large in our national spirit. The ideals of forward momentum, of economic growth and bounty, of stewarding a more righteous system of governance—these are all highly American and highly linked to the notion that we can engineer and will our way into a more perfect existence.

You might notice that these shifts are, if not explicitly religious in nature, closely tied in with the Protestant beliefs that dominated early America. It may be a stretch to call optimization a gospel and its physical manifestations, like the desert 777s, a shrine. Optimization has a shallower history, and its believers would hardly identify themselves as such. What optimization shares with religious belief, however, is a processing of the material world according to shared code and custom. It is also a set of practices: going faster, packing in more, saving money, individual retirement accounts, increasing productivity, keeping up with the Joneses.

The desert planes were designed and sold in an era of increasing long-distance flights, as the most optimal way to carry a large number of passengers across continents and oceans. The pandemic put a dent in long-haul travel and piggybacked on a broader trend toward smaller aircraft and regional hubs. Airlines began quietly canceling orders for wide-body jets. The planes in the desert boneyards will be gutted and sold for scrap. Time will tell what happens to the newer 777s: in late summer 2022, one of Air New Zealand's planes was brought back into service.

The desert hides all kinds of things: old military bases, faded still lifes of trailer towns, gatherings of rich techies high on manufactured drugs, and what was formerly the largest dump in America. Just east of Los Angeles, the Puente Hills Landfill's mountains of trash piled up over years, its smell drifting downwind. The dump was closed in 2013, and the site is now covered in a thin layer of dirt. The idea is that years into the future, it will be turned into a park.

DOWN THE ROAD FROM VICTORVILLE AND PUENTE HILLS, MARIE Kondo is on fire.

Kondo and her interpreter descend from a black Dodge van to a sprawling ranch home in Los Angeles. Like a Shakespearean sprite, Kondo bounces onto the stage. She enters the home, hugs the hostess, and proceeds to gently but firmly coach the resident family with too much junk through a monthlong purge-plus-marital-therapy intensive. By month's end, the garage is clear, piles of clothing have been dispatched to the Goodwill or the dump, and the children are smiling. Even the family cat seems satisfied.

All of this is part of Kondo's popular Netflix show, *Tidying Up with Marie Kondo*, which first aired in 2019. Kondo saunters into complicated lives and filled-to-the-brim homes, and in just under twenty-five minutes, as the audience binges on before-after frames, the house undergoes an intense decluttering. Interviews are interspersed with brief instructional videos, staged on a tidy minimalist set somewhere, with Kondo fluttering like a hummingbird just above a neutral-toned sofa. She rings a crystal, lights a candle, sprays some essential oils into the room. Suddenly, the air lightens, the bad spirits are shooed out. In one scene, Kondo wakes up old books by gently taping their storage boxes. In another, she teaches the viewer to fold baby onesies.

"Folding clothes is so much fun!" she exclaims, with utmost delight and calm.

The moral is clear: it's not merely magic; it is life-changing magic. It's right there in the name of Kondo's book, *The Life-Changing Magic of Tidying Up*. By harnessing your material reality, you can be saved. Kondo's trademarked system, called KonMari, is more than a way of organizing your home. It's a way of touching the divine. There is salvation in efficiency, and it's accessible to all.

Kondo's book was an instant bestseller when it was published in the United States, in 2014. A year later, Kondo was named one of *Time*'s one hundred most influential people. Why did her minimalism take off so spectacularly in modern-day America? She tapped into several of our anxieties. The most obvious one: a growing discontent with all our stuff, and a desire to contain it.

This urge toward minimalism, back when we had less stuff, went by a different name: frugality. One of Kondo's predecessors in promoting this virtue was no less fiery and just as organized. Benjamin Franklin was a writer, politician, and polymath. Frugality was one of thirteen virtues he wrote down when he was twenty years old to remind himself to embody daily. Among the others were temperance, tranquility, and order, which he described as "let all your things have their places; let each part of your business have its time."

Frugality, too, had its concise instructions: "Make no expense but to do good to others or yourself; i.e., waste nothing." Franklin kept a weekly schedule for practicing just one virtue at a time, noting that it was foolhardy to attempt to practice all thirteen at once. Indeed, he saw the process of working toward these virtues as a lifelong quest and a good unto itself.

These days, when we hear the word *efficiency*, we might picture a shiny steel machine, hard at work, or an office buzzing with meetings and conference calls. We might think of a government that runs well, a city in which no one goes hungry, a train speeding along the tracks, or a market in which everyone pays a fair price and not a penny more.

But the idea of efficiency, and its close cousin frugality, has far more quotidian roots. *Frugal* derives from the middle French *frux*, meaning "fruit" or "that which bears fruit." As the word evolved, the idea of producing, of being productive, also transformed from an agrarian context to a mercantile one.

Prior to the nineteenth century, frugality was a more common idea than that of efficiency (and optimization, which only later made its way into popular parlance). It was a household matter and a moral act. Waste not, want not. Plato considered frugality a virtue: "Live content with little." Even Warren Buffett, a public capitalist and private frugalist, hearkens back to an older age in advising: "Do not save what is left after spending, but spend what is left after saving."

This notion took particular root in early America, as Franklin's diaries bear testament. The Puritan colonists, stringing by to survive on a new continent, found opulence more than a little gaudy. North America was particularly suited to fostering optimization's ideal of control because it was mythologized, by European settlers, as both a blank slate and a wildness to be tamed. Work, prudence, and thrift were virtues that went hand in hand: "He that can earn ten shillings a day by his labor, and goes abroad, or sits idle, one half of that day, though he spends but sixpence during his diversion or idleness, ought not to reckon that the only expense; he has really spent, or rather thrown away, five shillings besides," wrote Franklin.

When it comes to frugality in particular, Franklin's words are tinged with moral judgment. Idleness is bad. Throwing away one's time, one's abilities, is the mark of sloth and sinfulness. At the same time, Franklin begins to couch frugality in the dollar-denominated language of modern efficiency. The language of shillings and pence. Frugality is an individual virtue and the way toward a prosperous society.

The combination of American self-sufficiency ("every man his own priest") and the need for structured answers about one's fate helped forged the frugality that Franklin embodies and markets to his readers. We began to seek through our work to organize and contain and engineer. Early colonists thus partook of a very Protestant solution—hard work, individual action, puritanical judgment—to a very American spiritual worry.

Yet while frugality gave a road map for individual action, it connected only tangentially, through Franklin's shillings and pence, to the whole. You could work hard and save, but there was still the nagging question: Were you right? This wanting to know, this need for certainty, would be at the heart of how frugality would transform into the culture of efficiency we see today.

In contrast to frugality, the word *efficiency* has its roots in the Latin *fic* (to make) and *ex* (out of)—to make out of. To be frugal is to fruit, like a tree. To be efficient, by contrast, is to make something out of what you have, to engineer, to build.

And it was a group of British thinkers who helped engineer a way out of the anxiety, by placing individual frugality in service of perfecting the greater good.

NINETEENTH-CENTURY ENGLAND WAS A TIME OF MOVEMENT AND change, as an agrarian society turned industrial. Trade was reaching out its arms across oceans, and for many people, the pace felt hard to understand.

An attempt to tamp this uncertainty was seen in a set of legislation, called the Corn Laws, aimed at protecting local agricultural commodities markets from global trade. The Corn Laws placed heavy tariffs and restrictions on imported grain, much as today's sugar laws do for Dakotas beets and Florida cane. (Corn, at the time, referred to a number of grains, including wheat, oats, and barley).

Somewhat suddenly, sentiment reversed, and the tariffs were repealed in 1846, in large part because of the efforts of a young man named John Stuart Mill. Mill argued that doing away with the Corn Laws would bring down the price of food and thus benefit especially the working class, for whom food was a huge expense relative to earnings.

Instead of simply campaigning to return to the past, Mill channeled public anxiety about industrialization around a goal. His was a belief that better engineering, rather than a retreat to older ways, would improve society. Mill wrote that the "spirit of the age" had changed such that "men insisted upon being governed in a new way." Indeed, the times were "pregnant with change."

With the repeal of the Corn Laws, food prices fell, and free trade made a comeback—as did the abstraction of tangible commodities, for centuries primarily bought and sold by those close to the land, into something increasingly traded by merchants and speculators.

The change Mill was speaking about ran deeper than trade. He is credited as one of the key thinkers behind utilitarianism, a philosophy that asks and seeks to answer the question, What is the greatest good, and how do we achieve it? Utilitarianism suggests that goodness can be measured and counted up—and thus, by extension, optimized. The actions that produce the greatest good, for the individual and for society, are also the most morally virtuous.

If the first step in optimization's growth was to democratize access to knowing, to link individual action to moral virtue, the second was to link individual action to the whole. Utilitarianism depends on the key notion that we, individually and collectively, are imperfect but eminently perfectible.

James Mill, the father of John Stuart, was a man of few words and bland ones at that. Or so claims his contemporary Thomas Macauley, who describes Mill's writing as partaking of "Quakerly plainness . . . cynical negligence and impurity of style." His style was, in other words, efficient. Mill's dour view on humanity led him, along with Jeremy Bentham, to first coin the idea of utility, writing that men must wrest from the natural world "the scanty materials of happiness."

From this vantage point, it was the role of society to civilize the barbaric and the "natural." The business of government was "to increase to the utmost the pleasures, and diminish to the utmost the pains, which men derive from one another." This view represented a major shift, a turn of attention from ensuring safety and salvation to perfecting life on earth.

Many of our most powerful institutions today have a debt to this strain of thought: from McKinsey consultants spreadsheeting their way to corporate perfection, to Mormon patriarchs shaping the material world as a display of spiritual prowess, to industrial economists and Harvard Business School gurus with their magic matrices and quadrants of success, and perhaps most of all to our political leaders. When Margaret Thatcher announced in 1981 that "economics are the method: the object is to change the heart and soul," it was clear what she meant. The economy was a tool that could be harnessed toward greater social good, toward changing behavior.

By transforming this anxiety about salvation and right action into action itself, Mill and his colleagues helped forge the lens of optimization. The individual was now seen as responsible not just for his own virtue but also for the reverberation of his actions in the world at large.

At the same time as it expanded, utilitarianism created a vacuum. When we seek to answer questions like, "Am I doing the right thing?," we have a limited tool set at our disposal. What is the right thing, anyway? Is it the thing that will be best for me, my family, or my community? How do I define best? Is it the thing that other people think is right? Which other people? How should I weigh my city's notion of rightness against my country's? The question quickly devolves into an anxious morass.

No longer is what's right so simple and well-known. Every decision, every question of rightness, requires a new interrogation of data, a new calibration. This is clear, for example, in American case law, lifted from the British system and commoditized and innovated in a way only Americans could do: precedent is built up, everything is up for debate, and all the more so if you can afford an expensive lawyer. This doesn't, of course, mean that right doesn't exist or is entirely subjective. It simply means that in any given moment, what Supreme Court Justice Oliver Wendell Holmes called the "marketplace of ideas," or sometimes simply the market, decides what's right. And then every individual decides for himself whether to buy it.

As efficiency grew as a mindset, its shortcomings were increasingly glossed over. The repeal of the Corn Laws initially lowered grain prices in England. By the late 1800s, however, new rail and steamboat lines made importing grain from North America and Russia much cheaper, and prices from these places were lower due to better soils and lower labor costs. By the early twentieth century, Britain had again imposed tariffs to protect its own farmers.

There was still a hitch with efficiency, as the turn back toward protectionism in England shows. It was all well and good to posit that society was perfectible, that individual action was tied to the greater good. When it came to doing good by God or by one's family, the path was clear. But how, in practice, were we supposed to know how to abstract this to things like international trade or inequality between classes?

While America also doubled down on protectionism in the nineteenth century, it was coupled with a westward expansion and can-do spirit of laws and customs yet to be written. Perhaps for this spirit, the idea of one's individual actions mattering was easier to square. There was forward motion. Regardless of the cause, the effect was that the third shift, the embedding and codification of this new way of seeing, happened first and largely on American soil.

Marie Kondo's gospel, as demure as it may seem, has at its heart an engineer's mentality. Its promises are also decidedly romantic. By harnessing the chaos of everyday life, its disciples aspire to a retreat from modernity, with its clanking wheels and too much stuff. By discarding the modern world's trappings, a sort of purification by fire, Kondoites are granted the return to an earlier, simpler age. At the same time, its systematic approach is grandiose and futuristic. Tidying up is the righteous path to mastery over the material world and, ultimately, over one's anxiety around it.

The ideals of utilitarianism, and the laissez-faire economics on its heels (economists adopted the term "utility" with great gusto), went hand in hand with the techniques of optimization. At the core of both is an idealism about top-down structure serving to bring about bottom-up good.

Utilitarianism and its minimalist offspring gave a locus to our anxiety about how our actions relate to the greater good, while only partially alleviating it. There was still the question of certainty. How could we know we'd exhausted all the possibilities? How could we be sure what we were doing was really the best?

STAN ULAM KNEW HE WAS MOVING TO NEW MEXICO, BUT HE DIDN'T know exactly why. So he went to the library to find out.

Ulam was a Polish-born mathematician—and later physicist— who first came to the United States in the late 1930s. In 1943, after Ulam had obtained American citizenship and a job at the University of Wisconsin, his colleague John von Neumann invited him to work on a secret project. All von Neumann could reveal about the project was that it would involve relocating, along with his family, to New Mexico.

So Ulam checked out a book on New Mexico. Instead of skipping to the section about the state's history or culture or climate, he turned to the opening flap, where the names of the book's previous borrowers were listed.

This list was a curious one. It happened to include the names of fellow physicists, many of whom Ulam knew, and many of whom had mysteriously disappeared from their university posts in prior months. Ulam then cross-referenced the scientists' names with their fields of specialty and was able to make an educated guess at the nature of the secret project.

Indeed, with World War II under way, Ulam had been invited to Los Alamos, New Mexico, to work on what would come to be known as the Manhattan Project.

The road into Los Alamos is precipitous, winding upward around the narrow edge of a cliff and into the small plateau upon which the town is perched. It's an appropriate approach for an alamo, or fort.

As you reach the top of the road, you first pass the small airport on the right and then a neatly organized little town. It's almost as if Marie Kondo has done her magic here, decluttering Central Avenue such that only the essentials remain. There are a couple small grocery stores, a florist, a gym, a pet food shop, and gas stations. Los Alamos has a few dozen churches and a fraction as many drinking holes. For a long while, the town had only a single bar.

The real heart of Los Alamos, however, is the highly secured lab farther down the road. These days, many of the lab's several thousand researchers and staff commute in from Santa Fe or elsewhere, winding up that narrow byway in the morning, down in the afternoon. Still, it's one of the highest-income zip codes in New Mexico. Supercomputers. Megascientists. After 7:00 on a summer night, the town is quiet but for the wind.

Seventy-five years ago, the contrast was even more stark, and the lab felt like not just an intellectual but a physical frontier. Ulam's wife, Françoise, described her impressions upon first arriving in Los Alamos: "[It] was a strange combination of Swiss village, construction site, and army post with PX and commissary where the sun always seemed to shine whatever the season. . . . Life was rustic and egalitarian."

There was, indeed, something egalitarian about this whole period, at least on the surface. The atmosphere at the lab was one of collaborative fraternity. The Manhattan Project itself was seen as a triumph of American ingenuity and scientific collaboration, even as it left pockmarks on the face of the earth. It destroyed cities, ended a war, and embedded the new prospect of nuclear annihilation. And then, postwar America saw one of the highest rates of growth, with relatively low inequality and inflation. Marriage rates were high. World war was over, or at least on hold. It was a time of economic stability.

Françoise added: "In retrospect I think that we were all a little light-headed from the altitude."

It was in this postwar aftermath that Ulam would make his most important contribution to the field of optimization. He and his family decamped from Los Alamos to the University of Southern California, where in 1946 he fell ill with encephalitis. It was a difficult illness, and while Ulam recuperated in bed, he kept busy with a deck of cards and game after game of solitaire. It was in these games that an idea was born.

As he laid out cards, Ulam wondered, What are my odds of winning this round? If he played enough times and kept track of the cards, he'd have data to describe his chances of winning. He could calculate, for example, which beginning sequences were most likely to lead to a win. The more games he played, the better this data would become. And instead of actually playing a large number of games, he could run a simulation that would eventually come to approximate the distribution of all possible outcomes.

When Ulam recovered from his illness and returned to work, he began to think about applications, beyond games of solitaire, for this method of sampling. A number of questions in physics could benefit, he surmised, from the diffusion of particles to problems in cryptography. A Los Alamos colleague with whom he still corresponded,

Nick Metropolis, had often heard Ulam refer to an uncle with a gambling problem. Because Ulam had conceived of the idea while playing cards, Metropolis settled on a code name, echoing the uncle's frequent adieus as he made for the casino: "I'm going to Monte Carlo." The method became known as the Monte Carlo method.

One of the first applications was a statistical approach to understanding neutron diffusion. This was a difficult problem because neutron behavior is not deterministic and thus is hard to predict even with computational aid. In a letter to von Neumann, Ulam proposed tracing the history of a neutron as it collided with others. Random numbers could be used to determine these outcomes. Using this method, a slew of interactions could be traced over and over, yielding a map of how a fission weapon might explode. A very early supercomputer, initially used to compute artillery firing tables, was commissioned to try it out.

The Monte Carlo method has come to underlie techniques in many areas of mathematics, physics, and machine learning. It combines the three ingredients of optimization: atomizing the computations, abstracting them to a model or simulation, and automating the computation to arrive at a solution. Taken from the realm of games and co-opting defense budgets, Monte Carlo quickly became one of the first examples of what today is par for a computer's course: performing large-scale simulations to understand natural phenomena.

Two of Ulam's colleagues describe the process of doing exact computations, prior to the development of the Monte Carlo method, to detail how the nuclear reactions in Edward Teller's proposed superbomb would play out:

[It] was tedious and exacting. While each step of the computation was understood, the complex interplay among the many components involved made the whole calculation extremely difficult . . . Since each step in the calculation depended on the previous work, it was necessary to complete each stage virtually without error . . . It is hard to imagine today that these calculations were performed with slide rules and old-fashioned manually operated mechanical desk calculators.

A set of methods developed around the same time, called linear programming, allowed for computational solutions to simultaneous equations. Returning to our lemonade example from chapter 1, this might entail writing out a set of relationships between proportions of ingredients and the drink's ultimate sweetness. Between the lemonade's sweetness and its sales. Between the season and the cost of lemons. These equations could then be combined and solved to find the recipe to maximize lemonade revenues at a given time of year.

The applications of these kinds of computational methods have reached far beyond lemonade. For manufacturing, they've allowed the shape of auto parts to be determined analytically instead of with time-consuming experiments. For transportation, schedules can be created, and updated, with the click of a button. For energy, it has meant the grid can be optimized to send power where it's needed.

Ulam's work on Monte Carlo marked a sort of golden age for the techniques of optimization. Colleagues such as von Neumann would go on to expand Ulam's idea, employing iteration toward the mathematics of dynamic programming for optimization. US Air Force mathematician George Dantzig soon developed what's now known as the simplex method, a systematic way of testing various possible optima within a set of simultaneous equations, in order to find the best.

Most of all, these technologies changed our way of looking at the world. Ulam's innovation gave us a map for connecting individual action to the scope of possible outcomes. It placed bounds on the probabilities once relegated to the furies and whims of fate.

Stan Ulam was an unlikely candidate to box in uncertainty in this way. As a scientist, he was expansive and collaborative; as a person, curious and kind. Ulam's colleague Mark Kac recalled that "Stan . . . came from a culture based on leisure and discourse . . . where you were in cafes all hours of the day or night drawing diagrams on small pieces of paper." As such, as Ulam's wife corroborates, he was "not very well suited for the American system of 'timetables' . . . He never became a nine-to-five man."

In his autobiography, Ulam expresses awe at the power of mathematics to make its mark on the material world: "It is still an unending source of surprise for me to see how a few scribbles on a blackboard . . . could change the course of human affairs." Yet Ulam's sense of awe is paired with dismay. His work on the Manhattan Project opened a new specter of nuclear destruction, and there remained a tension between his café-dwelling personality and the world he'd helped invent—and a question, about the extent to which human affairs or the material world should be changed by these engineered lines. Ulam's tension is not dissimilar from our national tension: on the one side, a love of engineered utopias; on the other, a fear of tyrannical control.

In the far-flung reaches of New Mexico, there were echoes of nineteenth-century Britain. If Mill and others delineated concrete goals and possibilities, Ulam turned that landscape of possibility into a set of probabilities, of concrete outcomes. In the quest to harness nature into a bomb that had the potential to end humanity's most fierce war, not to mention humanity as a whole, a bunch of scientists had invented a new method of computational control. It offered a means of connecting individual action to the collective.

If Ulam helped answer the how, we still faced the what. Now that I have the map, what are the best actions to take; what is the best route to follow?

MARIE KONDO DESCRIBES HER REBIRTH, AS A HIGH SCHOOLER, INTO an organization guru: "I was obsessed with what I could throw away. One day, I had a kind of nervous breakdown and fainted. I was unconscious for two hours. When I came to, I heard a mysterious voice, like some god of tidying telling me to look at my things more closely. And I realized my mistake: I was only looking for things to throw out. What I should be doing is finding the things I want to keep. Identifying the things that make you happy: that is the work of tidying."

The trademarked KonMari process involves looking at the objects—and behaviors, habits, rituals, relationships—that spark joy in your life, and saying goodbye to those that don't, all while expressing gratitude toward those you decide to keep. Says Kondo: "That, to me, is different from simply organizing."

After her high school epiphany, Kondo served five years as a *miko*, or shrine maiden, at a Shinto temple. Her duties included everything from keeping the space clean to performing dances and rituals in the shamanic tradition. Kondo talks of the experience in an interview with *Fast Company* and how it helped give shape to her worldview: "I think the shrine was a natural match for me because the foundation of Shinto expresses gratitude toward inanimate objects."

Indeed, an important ritual in Kondo's now-trademarked process of Tidying Up is to thank objects that have served their owner but that are no longer needed. Every object is infused with a life force, or *kama*, and to acknowledge its utility is to respect its life.

Watching the Marie Kondo system for folding clothes into a drawer evokes a sort of rapture. Everything fits into its place. There is no space wasted and no excess. It's optimal.

As the language of frugality morphed into that of efficiency, it took on both an increasingly individual and outward-looking lens. As access to knowledge was decentralized from priestly authority to individual reading of scripture, no longer did you have to trek miles to a church to catch glimmers of the divine. The American colonies provided the petri dish for these heretofore inchoate ideals to breed; Tocqueville writes, "[This philosophy] could not be followed except in centuries when social conditions had become more or less the same and men more or less alike."

Efficiency was on an upswing even before Stan Ulam, as the engineers and managers of the early twentieth century sought ways to codify their processes. Everyone from president (and engineer) Herbert Hoover to John D. Rockefeller sought to do more with less. Per a biographer, "Rockefeller's soul was more shocked and appalled by the inefficiency and waste of business than by the plight of the occasional struggler who fell by the wayside as his juggernaut moved forward."

Businesses were keen on these ideas. In the 1920s, Chicago-based advertising firm Mather and Company produced a set of motivational posters for factory workers, with peppy slogans advising them to "Stop the Leaks" in business efficiency. Mather claimed that each poster could avert a "10 percent leak per worker per day." A management system invented by Charles Bedaux proposed that workers be paid based on the number of work units they completed in a given amount of time and became the foundation of compensation at companies from Eastman Kodak to DuPont. Work by W. Edwards Deming, an American statistician and management consultant, was influential in establishing the Toyota Production System, a streamlined way of manufacturing that was then promulgated across thousands of businesses in every industry, under glossy headers such as Lean Manufacturing and Six Sigma.

Marie Kondo's philosophy joins together a number of these threads. The individual impulse to tinker joins with an appreciation for animate objects. In so doing, her teachings turn Ulam's map of engineered possibility into concrete action. Kondo's project takes a page from Franklin's frugality: individual virtue through right action.

By 2020, Marie Kondo's petite omnipresence extended out of the home, making its way into the workplace. The $12 billion Japanese e-commerce conglomerate Rakuten had acquired a majority stake in KonMari Media. Kondo coauthored a book with a business-school professor titled *Joy at Work: Organizing Your Professional Life*. In the acknowledgments section, she states her mission: "Our company vision is to organize the world."

As with her home-decluttering program, the book offers a set of instructional ideas in an easy-to-follow list. There are chapters devoted to everything from tidying your workspace to "Tidying Time" and "Tidying Decisions." Readers are asked to look soberly at each work task and keep or discard it based on criteria, such as "Is this task required for me to keep—and excel—at my job?" or "Does this task contribute to more satisfaction at work?"

There are also more basic recommendations, such as "Beware the Scan Trap" and "Rule 2: Store your papers upright." The reasoning for the latter is dutifully explained: "For optimal efficiency it's crucial to store your papers in a hanging-file system . . . Storing them this way makes it easy for you to see how many papers you have. It also looks neat and tidy."

When we speak of efficiency today, we mean a very different thing than was meant in 1720 or even 1920. Take repairing something that's worn out. These days, there's still some small virtue in darning a sock or reupholstering a chair, although fewer modern Americans take the time do so or would call it efficient. Baking a pie at home, rather than buying one at the store, is something we might do for entertainment rather than out of economic necessity. In fact, because of the efficiencies baked in (so to speak), a store-bought pie is often cheaper. Likewise, knitting a pair of socks as a gift (albeit with expensive yarn bought online and shipped thousands of miles) is now seen as a thoughtful gesture and sign of leisure. But we'd still say the most efficient thing, in terms of both time and money, is heading to Costco for a sixteen-count box of handheld apple pies from the freezer section or an eight-pack of Kirkland-brand crew socks.

As our language shifted from a virtue to dollars or time saved, the role of saving changed, too. Whereas frugality is about the intrinsic virtue of an action—baking a pie is frugal unto itself—efficiency asks about the system as a whole: Is baking a pie or buying one at the store more efficient?

It also took on an economic tenor. William Stanley Jevons, whom we'll meet in full in the next chapter, proposed a new theory about the value of commodities. Prior to Jevons, in the early 1870s, the dominant theory to explain a commodity's price derived from Adam Smith. The labor theory of value posited that commodity prices reflected the labor put into them. This included both direct labor, such as planting and harvesting a crop, and the labor that went into the necessary capital, such as the harvesting equipment used by farmers.

Jevons flipped this idea on its head. Instead of a commodity's value being measured by what went into it, he proposed that its value lies in what it is worth on the market. That is, in the utility it provides to those who buy or use it.

This flip had major consequences. Jevons's idea opened up the possibility of seeing a price, and the market as a whole, as a set of individuals determining worth and collectively maximizing utility. It turned the concept of economic value from something additive into a closed market or system that, like lemonade, could be optimized.

STAN ULAM RETURNED TO LOS ALAMOS ONCE MORE, AFTER HIS RE-tirement from academia in the 1970s. This move was different. The first time he'd decamped to New Mexico, he'd joined an elite if ragtag group of scientific minds working on a moon shot. No one knew if the Manhattan Project would succeed. The second time, it was after the project's success and the official opening of the Los Alamos National Lab, after Ulam's illness had led him to devise this Monte Carlo method.

This time, Ulam returned to a different legacy. By 1960, *Time* magazine had named "US Scientists" the person of the year. The American economy barreled forward, fueled by the promise of scientific collaboration and innovation. Growth seemed unlimited. And the National Lab was where it would all go down.

Since Ulam's initial stint in New Mexico, much of America had come to believe that engineering and computational technologies would be central to the coming decades. Indeed, the ideas first implemented in national labs like Los Alamos were copied in years that followed: by Wall Street in the 1980s, by Silicon Valley at the dawn of the twenty-first century, by the biotechnology work of recent years.

Yet if computational methods allowed the modern practice of science to rise, it wasn't without its shadow. In his autobiography and his later talks, Ulam looks back with mixed feelings on the era he helped create. After he and Françoise settled in Santa Fe, Ulam drove up to Los Alamos to give a Sunday talk to the congregation at the Unitarian Church. He recalls that "the discussion centered on . . . the relation of science to morality." It was something another mathematician, Henri Poincaré, had considered in his 1910 writings. Ulam cites Poincaré, and the tremendous changes since. "The questions were less disturbing then. Now, the release of nuclear energy and the possibility of gene manipulation have complicated the problems enormously."

In his later years, Ulam's easeful optimism had softened. Perhaps not everything was meant to be harnessed or explained.

DOWN THE ROAD FROM WHERE ULAM GAVE HIS SUNDAY TALK, BARbara Grothus is disassembling a different kind of church. Barbara's father, Ed Grothus, was a machinist at Los Alamos National Lab for twenty years. Born in Iowa, he'd joined the merchant marines, after which he worked for a while on cruise ships before settling down in New Mexico. The lab offered him a job in 1949. In the late 1960s, Ed's political views began to shift, and he took to protesting war and nuclear proliferation. A decade later, he bought and converted the Piggly Wiggly grocery store in Los Alamos into an antinuke shrine featuring objects from his time at the lab, as well as assorted atomic gadgetry. Two large trailers housed these sundry items, and those that didn't fit inside, such as a pair of granite obelisks—"doomsday stones," as they were called—were placed outside. Next door to the grocery store was an old A-frame church, which Ed transformed into the First Church of High Technology. Quite naturally, he minted himself minister and, later, pope.

On the anniversaries of Hiroshima and Nagasaki, Ed would display a large banner outside: "We apologize for the nuclear bomb." Of his former employer, he said he admired the work in medical research and other fields but that "it's been a bad business. They should leave the uranium in the ground."

After Ed died in 2009, Barbara kept the site alive for a handful of years; when I lived briefly in Santa Fe, one could still visit and perhaps purchase a souvenir to take home. But the energy and disciples that Ed had attracted had dissipated, and Barbara eventually sold off the materials and shut down the museum.

When asked recently by a reporter, "War. What is it good for?" Barbara Grothus replied: "It has been good for the New Mexico economy. Los Alamos has a big economic footprint. It impacts other industries and it's been very good for certain people in our state."

Barbara's answer was half-ironic, half-searching for a silver lining. For the most part, she found the whole affair of Los Alamos "depressing," and when asked what should become of the former museum site, she quipped, "It would make a good grocery store."

Not everyone is content with the gospel of optimization and its accompanying belief in a world that's ours for the taking. While Ed Grothus's church attracted only a small following, it stood against the notion of progress at all costs; in the next chapters, we'll see others. Although the property has sold, remnants remain on display in the Bradbury Science Museum in Los Alamos.

IN 2019, MARIE KONDO MOVED TO LOS ANGELES. HER THIRD CHILD was born in the spring of 2021. The whole family can be seen beaming in the Southern California sunshine to four million followers on Instagram.

Kondo's story is perhaps less linear than it appears at first glance. Her philosophy didn't emerge in a single vision, although in retrospect Kondo attributes its inspiration to her time serving at the

Shinto shrine. Nor was bestseller status an entirely unexpected outcome of her book. She'd entered a book-writing competition and was noticed by a major self-help editor who saw the makings of a guru. Kondo dedicated herself to fleshing out the book and grew into the part.

That Kondo is a complicated character is lost somewhat in the neutral tones and folded baby socks of her messaging. A complicated savior is somehow less appealing.

Around the birth of her third child, Kondo realized she was being pulled in too many directions. She describes in an interview, true as ever to her messaging: "Since I had that epiphany, I've made it a point to prioritize time for joy in my life, especially when I'm busy. I deliberately schedule in time for things I enjoy or want to do."

Kondo is a product of our times and a poster child. She's an optimizer par excellence. But she's also a true believer who, by organizing, has brought joy and aesthetic appreciation back into our impoverished language.

It's hard not to wonder where the Los Angeles ranch-house family is today, after Kondo helped them declutter on Netflix. Have they reverted to their old ways, the same old patterns, filled their house with new junk? Have they swung the opposite direction, into full-on minimalism? Somewhere in between?

For many of us, it's become harder to trust the gospel. It no longer squares with our daily experience, with the cheap junk hidden away in our own garages, with the growing debt, with the mixed-up messages on Twitter and TV. We went looking for answers about our purpose on earth, and the likes of Franklin, Mill, Ulam, and Kondo offered partial reprieve. Their ideas connected our actions to society at large and cast the material world as one we could improve. So we tidied our homes, we built jets and carved up mountains and divided the atom. And now if we go looking for them, we'll find the ruins hidden away in airplane graveyards and beneath thin layers of dirt, in the blinding light of the desert.

4

METAPHORIC BREAKDOWN

IT CAME OUT OF THE BLUE.

In February 2021, half of Texas shut down. The polar vortex, a low-pressure zone of cold air, reached its icy tentacles southward, battering the southeast with winter storms for nearly a week. Temperatures crawled below zero, hitting records for the century and blanketing normally balmy latitudes in snow and ice.

Households turned up the heat just as various power suppliers were going down. The Texas electric grid partially collapsed and was on the verge of shutting off entirely. Some forty-six thousand megawatts of power had to be taken offline, equal to the consumption of nine million homes on a high-demand day. Natural-gas pipelines froze, valves were iced shut, and windmills stopped turning. When natural gas is cut, gas-fired power plants can't generate as much electricity. And natural gas, which heats about 35 percent of homes in Texas, relies on electricity to be processed.

The grid reacted with rolling blackouts. At the peak of the crisis, nearly five million customers across the state were without power, and many more faced intermittent blackouts. The streets in Austin were frozen. Boil-water orders were put in place after power was cut to water-treatment plants. Houston residents burned charcoal and scrap wood to keep warm and lined up to fill water buckets from public spigots. Near South Padre Island, rescued baby sea turtles were warmed in makeshift indoor pools.

On top of that, because of extremely low supply and high demand, the price per megawatt hour for electricity jumped from around $30 to around $9,000. As Texans emerged from the cold and the bills rolled in, many were astounded to find they owed several thousand dollars, even as their homes remained frozen.

The Texas disaster provided a sort of release. An external, visible, feelable disaster. The lights were out. The house was cold, fingers were numb, water boiled on the stove. Supermarket shelves were empty, and cars spun out in the icy streets. Families huddled in blanket forts and ate cold meals by candlelight. Finally, after distant wars and unseen viruses, this apocalypse followed the script. It made sense.

We saw in the previous chapter a malaise, rooted in the disconnect between the promise and reality of optimization. Perhaps Mike Campbell describes the feeling best. Mike is an unruly war veteran making his way through Ernest Hemingway's novel *The Sun Also Rises*. His hard drinking earns him a handful of enemies, a mountain of stories, and a whole lot of debt. When asked how bankruptcy befell him, he replies succinctly: "Two ways. First gradually, then suddenly."

Leading up to the financial crisis of 2008, the defaults on Florida ranch homes trickled in slowly at first, as the credit-default swaps built up. Decades of poor fire management compounded gradually in California, along with suburban homes encroaching bit by bit into fuel-filled forest. And no one paid much heed to the small fractures in supply chains building over the early years of this century, until a pandemic hit, when they seemed suddenly to break down.

Gradually, there is a buildup of abstractions and automation, of dry fuel and overextended loans. Gradually, slack is lost and fragilities grow. Gradually, those in charge of a system lose track both of its design and of its vulnerabilities. Gradually, we sweep things under the rug.

And then, suddenly, the lights are off and the pipes frozen solid.

Texas's blackout was decades in the making. The electric grid is a complicated tangle of lines and plants sprawled across the North American continent, a mishmash of power generators, operators, and consumers connected by underground lines and allocation algorithms. The US grid has three parts. The Eastern grid stretches nearly halfway across the country, to the foothills of the Rocky Mountains. The Western grid runs the rest of the way to the Pacific Ocean.

And then there's Texas. As in many things, and for various reasons, Texas stands alone. During World War II, the Texas Interconnected System was built to bring large amounts of power to the Gulf Coast and fuel the war effort, and the state ended up with its separate control of electricity.

Texas's uniqueness was both a buffer and a liability. Had the state been more connected in February 2021, and the winter storm more localized, the Eastern grid might've been able to step in and lend assistance. Here, however, the initial cold-front polar vortex was hitting much of the southern United States. There was no help available.

Moreover, a number of plants were down for winter maintenance. But maintenance and weather are things that can be described in advance. A rich engineering literature helps predict failure points with statistical accuracy. Like the movements that a seismograph picks up before a large earthquake, there is a "signature" to the conditions that precede a breakdown.

So why do they so often hit us as a surprise?

ALLEN GILMER HAS AN OPINION ABOUT WHAT HE CALLS THE TEXAS fiasco. He says he'll get to that in a moment. Right now, Allen's telling me about The Last Knife Fighter. A bearded buckaroo perpetually accompanied by a canteen of whiskey and prone to 3:00 a.m. drunken phone rants, TLKF is, according to Allen, the last real singer-songwriter alive today. He's an ex–hockey player, a hard drinker, a recluse, a rebel, a hero; a man in full. He lives on a few rotten acres outside Paradise, Texas, and calls himself the Father of Outlaw Folk.

TLKF may be many things, but he's no wallflower. He's made his songs available on YouTube and Spotify and Instagram. One of them is even sponsored by a whiskey company. He occasionally plays live shows for beer money and for kicks. But rest assured, says Allen, the guy is the real deal.

Allen and I are sitting on a sprawling patio in Austin, Texas, sipping margaritas. Or, rather, I'm sipping. He's here on the heels of a business lunch and is a couple drinks ahead of me. Allen is both something of a mellowed-out rebel himself and the former CEO of a half-billion-dollar company in the energy industry. And he's taken a liking to TLKF. The story verges on allegory, Allen rhapsodizes. No one else is left in country music who's not some Nashville sellout.

Another round of drinks arrives, and Allen continues. We've lost that sense of singular focus, not just in music but in everything—the singular focus that put a man on the moon. We have no more moon shots. Now it's all electric cars and subsidized windmills and other bullshit that's actually worse for the environment than oil.

I don't disagree with the spirit of Allen's point. There's a lot of green-energy fakery out there, masquerading as salvation. There's been more than one vanity rocket ride into "space." And for better or worse the days of scientists like Stan Ulam coming together to solve a problem with global consequences feel long gone. At the same time, I'm not sure TLKF is the first person I'd pick to pilot the next Manhattan Project, much less my moon-bound spaceship.

What if, I wonder, moon shots are in decline because the whole idea of optimizing our way out of things is in decline, and we don't know what to put in its place instead.

The singular focus of optimization gave rise to incredible innovation. That focus has also come with trade-offs: the loss of slack or downtime, the abstracting away of place and particulars. It's led to a loss of human scale, making events like the Texas-grid breakdown more frequent.

As these trade-offs come into focus, there's a tendency to sweep them under the rug. To pretend the bill for factory agriculture will never come due, in the form of depleted soils and a population that's lost the knowledge of how to farm. To imagine grid failures can be stopped by market mechanisms alone and conveniently forget that electric cars require combusting mountains of coal.

There's also a tendency to lionize either a past golden era, as Allen suggests, or a future utopia. In the next chapters, we'll look at ways in which people have aimed to solve for the things lost because of optimization. Here, we consider what it means to live through the point when we finally come to terms with its fragilities.

What is this point? What does it feel like? Perhaps, like Mike Campbell, we find ourselves in a state of perpetual low-grade bankruptcy. Optimizations themselves continue to function, especially at smaller scales. Trains stick to their schedules; factory lines churn on. What is in decline is a metaphoric era, the belief that more efficiency is always better. We see this first around the edges: a failure here, a lag there. And then the breakdown hits us suddenly, as entire industries seem brought to their knees.

The year 2008 was a moment of rupture for the global banking system. It took extreme intervention to keep major financial institutions afloat, institutions that were supposed to stand for the freedom of markets and the limitlessness of growth. By 2018, there was a crumbling of belief in major tech companies, although, as with the banks before them, most Americans continued to rely on their services. We're seeing a similar breakdown in the early 2020s, as the idea that global supply chains will always deliver and commodities will always be cheap goes bust. A metaphor in decline doesn't mean we've moved past it, only that there's now a collective revelation of its flaws.

The previous chapter traced the roots of this metaphoric era, from colonial America through the British political theorists of the nineteenth century, into the postwar innovations in computing, and up to Marie Kondo's minimalist movement. Could the roots of optimization run deeper still? Some argue that our modern era began with the creation of markets, or of money itself, or of tools to shape reality around us, or even with the adoption of the Christian story of the fall from Eden: when we bit from the apple of knowledge, when we used this knowledge to engineer Nature rather than submit to her fancies, we opened the door to all the disenchantments of modernity.

Because this book is concerned with the more recent story of optimization, and how the virtues of efficiency both seduced and failed us, we mark the bookends of the age more narrowly, beginning around the time of Newton, around when the early culture of America was fomented and Protestant ideas about individual and collective salvation were working their way into the modern consciousness.

And we can just as well start to mark its ending. If one of America's core myths was westward expansion, the closing of the frontier left an uneasiness, a feeling that everything was now mapped and enclosed, known, that a certain wildness had been lost. Indeed, many early American accounts describe the continent in Edenic terms, its domestication both the goal and the deflating ending. For as long as the frontier was open, growth and efficiency went hand in hand. When it was closed, efficiency needed to cannibalize to survive.

Optimization, after all, works best, most purely, when the boundary is closed, when the game is zero-sum. When you can't move pieces on or off the board, or trade with someone outside the match at hand. When a natural resource such as oil or gold seems limitless, unbounded, it's an explorer's game. Once the field is surveyed, the mine tapped, favor shifts to the managers, those who control the territory and know the map. For a long time, we ignored this shift, with the promise of never-ending resources, of infinite growth.

Of the early 2022 stock-market crash, investor Peter Atwater says:

"It's not a 'tech sell-off.' It is a massive repricing of abstraction. We had a bubble in dreams." A similar thing could be said of the mythos of efficiency. Our continued enthusiasm was driven in part by a faulty assumption, the dream that growth could be forever extracted from a bounded system.

The sun is going down, and the patio plants seem to droop, as Allen segues from country music to the future of American oil. It's May 2021. There's a feeling that we've emerged from a long winter and into a new world. Around the globe, the pandemic still rages, but early treatments are making inroads in America. The collective chatter has an upward inflection: Maybe this is finally behind us? The sun is warm, there's a gentle breeze, and everyone's bursting into bloom.

The best, the only, the first—it feels easy to make heroic claims in the springtime. Optimization was brought to life by Protestant scholars, British industrialists, an international cohort of scientists posted up in a high desert lab. But it found its surest foothold in late twentieth-century America.

And in the past couple of decades, that foothold has started to slip.

ALLEN IS THE FOUNDER OF DRILLINGINFO, NOW CALLED ENVERUS, A corporation that sells data to the energy industry. Its clients are grid operators and investors in oil and mineral rights. When he's finished telling me about TLKF, Allen is keen to absolve the oil and gas sector of undue guilt in the Texas blackout. They're blaming natural gas, he says, but really it was a system-wide breakdown initiating at the electrical power generation plants and leading to a cyclical effect.

To failures of both coal-fired plants and natural-gas providers. He points me to a report by Enverus, full of data and charts, showing not only prices across the region before, during, and after the storm but also the failure points in the grid and a breakdown of the losses in generating power experienced by different providers.

When Allen started out, the world looked very different from the one now captured in the graphs and charts. It was a world unmapped. He started the company out of a conversation with the colleagues who would become his cofounders, about how to value a piece of land. How do you determine, in expectation, how much value it will provide? How much oil might lie below? How much potential energy might it hold?

The founders brought to these questions the science of probabilistic analysis and a knowledge of geology. It meant combing through and combining disparate sources of data, in an era before digitized records, before the pervasiveness of computers or computing power. To date, the industry had relied primarily on wildcatters and whims, on brute-force drilling and good luck.

Our patio chatter continues as the waiters clear and reset tables. Allen starts telling me about a movie he's producing. But to tell me about that, he needs to rewind a bit further. All this oil stuff, he explains, actually began with a cowboy story.

A cowboy named Wil Andersen, in fact. Andersen is a frontiersman and cattle driver outside Bozeman, Montana. He's the character played by John Wayne in the 1972 movie *The Cowboys*. In the film, Anderson needs help with a four-hundred-mile-long cattle drive and is having trouble finding experienced men. He finally recruits a band of young schoolboys to help out.

Soon enough the drive gets going, long days shooing cattle up plateaus and along riverbeds and then making camp for the night. At dawn, the work begins anew. "We're burning daylight," yells Andersen, as he shuffles the boys out of bed and into their saddles.

And before too long they hit some obstacles. Trouble appears. Asa "Long Hair" Watts, played by Bruce Dern, is a rabble-rouser with an axe to grind. Watts and his gang had applied for jobs with Andersen, only to be rejected for their bad attitude. When they meet along the way, Watts homes in to take advantage.

The outlaws trail the old cowboy and young boys and the camp cook, aiming to steal their cattle and finish the drive themselves. When an armed Watts finally descends on them, Andersen knows he'll need to fight. Fearing for the safety of the boys, he tells them to protect themselves by proclaiming they're not involved. The dispute, he says, is between him and Watts. The boys nonetheless step up to help their patron. Watts and Allen get into a fistfight. Anderson appears to be winning, until Watts reaches for his gun.

In every John Wayne movie, there's a bad guy and a fight. There's good and evil, and squarely on the side of good is John Wayne, who battles for what's right and ultimately emerges victorious.

In *The Cowboys*, as Watts is losing the fistfight, he grabs his gun and shoots the unarmed Anderson once in the leg, then twice in the heart, before he and his men run off with the stolen herd of cattle. The boys and the camp cook rush to Andersen's aid, but it's too late. The boys are left without a leader.

It's the first time John Wayne has been killed on screen.

Sitting in the dark El Paso theater, next to his own father and a four-hour drive from home for the movie's premiere, is a thirteen-year-old boy named Allen Gilmer. His hero is dead. His world will never look quite the same.

ALLEN CAN BE FLIPPANT IN TELLING HIS OWN ORIGIN STORY, GUSSY-ing up the more macho parts, attributing his trajectory to luck and grit, making light of any intelligence or strategy that might have helped along the way. To an interviewer for an industry magazine, he describes his choice to study geology with a reckless gesture: "Geology is the biggest bait and switch major there is, because they sucker you in with fossils and cavemen and dinosaurs and really cool stuff like that."

When Allen and his colleagues first started Drillinginfo, the company aggregated permits and oil rights. They focused on Texas, then expanded into other geographies and industries. At first, this was a manual process, getting permits from the courthouse and railroad commission, then leasing activity from the courthouses all over the state, all 254 counties manually entered into computer databases. Allen recalls driving all day to county courthouses, from one end of Texas to the other. By connecting these pieces of information, they were able to construct a map of probabilities of what might be where, which they sold to assessors and investors.

At the time, the industry was very regional. "An oil man would just cover his region," Allen explains to me. "You didn't know the other places." Drillinginfo connected the dots and created the map.

When we make a map, it changes our view of the territory. When we describe a model, over time it becomes more cemented. We elect data sets, statistical tests, analyses. Soon we forget the messiness of exploration, leave behind our intimacy with the numbers.

So, too, as the world itself becomes mapped, we throw out our compass and sense of adventure, as well as our comfort with getting lost. We begin to use maps made by others. We forget the ferns underfoot. Before long, we've got the windows rolled up and a GPS barking instructions.

If Stan Ulam helped introduce a new age of computational modeling, Allen Gilmer took the concept and ran. The goal of an expedition, oil or otherwise, is to find the deepest deposits, the highest peaks. Algorithms were developed to optimize this search—as a textbook on algorithms will tell you, there are multiple ways to search a landscape or a set of files in a directory: depth first, breadth first. Once the map is known and the shortest path established, recalling the information becomes much simpler.

As the techniques of optimization expanded into bigger and bigger projects, discovery and maintenance began to separate. This happened in companies such as Google, it happened in scientific labs, and it happened in finance and logistics. Roles were atomized, models abstracted, the whole thing automated and managed by people who hadn't been involved in its creation. Inputs and outputs were packaged into black-box algorithms, in a way the clanking factory lines of the machine age could not.

George Lakoff and Mark Johnson, in *Metaphors We Live By*, argue that a metaphor structures and changes how we see and interpret the material world. Thomas Kuhn says something similar in *The Structure of Scientific Revolutions*. New ideas punch holes in old models, and novel ways of knowing emerge from the wreckage.

In Allen's world, optimizations and algorithms reduced open-ended exploration to game-theoretic calculations. The more quickly new carbon deposits could be found, with the least output of human and computer power, the better. The role of modeling shifted from an older-school idea of understanding to a newer concept of control.

Something else happens when the map is known, the tallest peaks identified. Some magic disappears. Over time, Allen explains, the modelers became less interested in the data. The people running the grid forgot how it was made. The people on the ground weren't trained in geology, nor did they have a knowledge of the land. Allen's world of pickup trucks and notepads and back roads was replaced by a slick new office in Austin.

Although he was never one himself, Allen identifies with the wildcatters and ne'er-do-well lifestyle of people such as TLKF. Allen's work changed how the oil and gas industry did exploration and valued land. Yet as he mapped the physical world, and profited greatly, it changed him. Like the farmers of the Dakotas who became as much commodities futures traders as harvesters of grain, the dirt under Allen's fingers disappeared. Eventually, he resigned from active management of the company, becoming its chairman. Part of him, I realize, feels dismayed by the civilized landscape he helped create. Like John Wayne's death, it doesn't quite compute.

THE DISCOVERY OF GOLD IN 1851 IN NEW SOUTH WALES, A WORLD apart from 2021 Texas, kicked off an Australian gold rush. Three years later, nineteen-year-old William Stanley Jevons (whom we met briefly in the previous chapter) abandoned his studies in London to take an assayer job down under. When Jevons finally returned to England in 1859, he fixed his academic mind on the impacts of gold rushes on price. And soon after, he took interest in a dirtier commodity. Coal was on people's minds: they fretted that as England's coal supply was exhausted, prices would skyrocket.

What happened was just the opposite. As the supply of coal dried up, engineers found more inventive ways to mine it, thus reducing the price of extraction per unit. And as coal became cheaper and more abundant, people invented new uses, becoming more reliant on the fuel in their homes and businesses. Jevons's book *The Coal Question* addresses the relationship between a resource's abundance and its price. The counterintuitive result is now referred to as Jevons's paradox: as mining a resource becomes more efficient, as the resource becomes more abundant, demand tends not to decrease as expected but to go up.

Jevons's paradox is echoed in a modern-day paradox of optimization. The better we get at optimizing locally, the more uses we find for optimizations generally—and the worse we get at reconsidering the idea of optimization itself. The farther Allen gets from the on-the-ground data and records pulled from courthouse archives, the more the map he's created replaces the terrain.

Because the Texas energy market, like the coal market in question with Jevons, is deregulated, prices emerge nearly at will. During normal times, this means a competitive market in which consumers pay for what they use at stable rates. But when supply plummets and demand rises, as it did in February 2021, prices can go through the roof. It's a system optimized to work fantastically, when everything's working fantastically.

How, in practice, does the grid "decide" where to allocate power and how to price it?

There are several kinds of algorithms at play. We can think of it like a three-layer cake, with icing on top. The bottom layer represents the network algorithms that decide how to allocate energy and what to do when new power plants, or customers, are added. In the middle are systems responsible for near-term planning. If a large office building that's normally shut down on weekends suddenly ramps up on a Saturday, where will this power come from? If Texas is hit by the polar vortex, how will the lights stay on?

The top layer serves to route communications between nodes. Here, pricing mechanisms act like a clearinghouse, accepting bids and offers and allocating power to different customers based on the prices they're willing to pay and receive.

And finally, there are financial instruments, layered like thick icing atop the cake, that allow power generators—and in some cases customers—to smooth their expected revenues over time, or to hedge against the kinds of spikes seen in Texas in 2021. This comes, of course, with a cost that's pocketed by the financiers.

The US electrical grid, and Texas's in particular, developed as a combination of local optimizations: existing power plants were strung together, efficiencies piled up. Auction mechanisms allowed for complex bidding and communication.

It was easy to assume that each component piece was optimized even as the fracture points emerged. Tearing the whole thing down was inconceivable: it would disrupt power for millions of residences, businesses, and industrial customers. Instead, the grid was asked to bear even more demands. Not just more output but also new types of fuels and different usage patterns.

For millions of freezing households, the breakdown was not what was supposed to happen. And yet, for those who designed the market mechanism, the grid functioned exactly as planned. There was huge demand in a time of shrunken supply. Prices adjusted automatically. "Only" 40 percent of the grid went down. According to Bill Hogan, professor of global energy policy at Harvard and "architect" of the Texas-grid algorithms, "This crisis situation was unprecedented . . . but the market design coupled with the swift action of the system operator helped prevent an even worse catastrophe."

Per Hogan, in times of scarcity, as was the case in February 2021, it's "especially important" to price correctly. That is, to let the algorithms do their thing. Even if it means $16,000 bills for a residential customer. Hogan's argument is that if consumers see the price signal, they will reduce demand in peak times. He concludes, "All this was necessary under the circumstances."

It was necessary, and according to those who designed it, even planned like this. So why aren't breakdowns such as Texas's at least predicted, and stopped, in advance?

Proponents of optimizations like to cite their impartiality and elegance. The auction system behind Google ads, for example, entails quite technical, and also quite beautiful, concepts. In the case of Texas, however, these mechanisms are knit closely with the physical world. And the physical world is patched with connections and fixes and what economists like to call "frictions." These physical interactions are far more difficult to measure and model theoretically than, say, a Google Ads campaign.

A similar issue is behind other crises, from energy and environmental to those in financial markets. All too often, models are overfit. Overfitting a model occurs when we rely too heavily on a small subset of real-world data to make an inference. The model then too closely reflects that data and fails to predict larger conditions very well. For example, given a table of February weather conditions in Florida, we might conclude: it hardly ever rains here, and it doesn't get too hot. We'd be ignoring, of course, precipitation numbers from hurricane season in the fall and the peak of hot temperatures in the summer.

If the conceptual model of optimization is overfit, once the spell is broken, we begin to see failures everywhere. The year 2022 was the worst on record for airline cancellations. A baby-formula shortage went on for months. Indeed, in various instances of breakdown, it's hard to know whom to blame or how to set it straight again. One month after the blackout, in March 2021, the container ship *Ever Given* hit high winds and ran aground in the Suez Canal. About a quarter mile long, the *Ever Given* blocked off the canal, leaving other ships helpless to pass. There it remained for six whole days. Each day ticked away tens of millions in cost to the global supply chain.

When the Suez opened in 1869, it became the first waterway to connect Europe and Asia, as well as the Pacific beyond. The canal offered not only a shorter route but a less treacherous one, avoiding the capes and allowing ships to sail anytime of the year. Before too long, larger ships were built, and trade began to galvanize around these optimizations.

However, the strong winds of the Suez made traversing difficult without power. So more fossil-fueled vessels were adopted. With the invention of the shipping container in the mid-twentieth century, ships could be loaded more readily with atomized cargo. At the time, it didn't seem like anything could interrupt this flow of progress.

AS IN LIFE, SO IN ART. BRUCE DERN'S KILLING JOHN WAYNE IN *THE Cowboys* was a milestone, although it didn't seem so then, as Wayne went on to star in many more films. A thread that had held a genre together was frayed.

John Wayne was the archetype of strong, controlled masculinity. He was beat by measly Bruce Dern, in front of young boys, no less. Those boys quickly grew up and finished the cattle drive in Wayne's stead. The promise of optimization was that perfect engineering would never break. The public breakdown of large-scale systems, such as the grid or supply chains, broke this promise, just as John Wayne's death broke a silent contract in Western films. Wayne kept on showing up on screen, just as the energy market continues forth in Texas. But everyone could now see the fray in the weave.

Often, we read about the decline of a culture or empire as something precipitated by a singular shock: the 1917 revolution that ended Czarist governance in Russia, the collapse of the Aztecs by the introduction of smallpox by Cortés and his crew. Likewise, we are wont to see metaphoric breakdown as resulting from a sudden change in zeitgeist, rather than a slow wearing thin of perspective.

Just as often, shifts in epistemology occur quite gradually, even if they are later denoted by a single event. As a child in 1989, I watched my mother and grandmother watch the TV, seemingly for days on end. I was young, but I remember the scene vividly, the black-and-white static, the broken plastic knob to change the channel, a big building and a crowd. The camera replayed footage. There was noise, there were angry shouts from the crowd, and there was a helicopter and the women crying quietly in front of me, half a world away. It had all happened suddenly, I later learned, but it had been brewing for a while. One small shout, a crowd that couldn't be controlled, and a dictator of nearly forty years shot on Christmas Day.

The regime they'd grown up with, in Romania, had fallen. There was relief but also uncertainty. For so long, they'd stood against something. Now they weren't sure where to stand.

Did Allen's heroic age ever exist? Did it persist as long and un-troubled as we'd like to believe? Or was it first built up as a slapdash movie set, becoming a fixture over centuries, until we could only assume it was real?

Allen founded his company when the oil and gas industry was in the pits, in part because he couldn't stand the idea of groveling for jobs with the "majors": the Shells and BPs and Exxons. Now, these same majors are clamoring to develop shale in the Permian Basin in west-ern Texas, an area until recently the province of independent explor-ers and producers. In just a few years, the big companies have spent on the order of $10 billion to buy land. Exxon wants to change "the way that game is played" with shale, per its CEO, Darren Woods.

Allen, meanwhile, has collected his take. But I wonder if he also lost a part of his world. By the time Drillinginfo rebranded itself Enverus in 2019, the company had grown to nearly 1,500 employees.

Most of what Allen tells me, and most of what he says in the magazine interviews I read, is focused on the past.

He says to a Reuters reporter, "Nothing is a bigger motivator than, 'Am I going to be alive tomorrow?'" And then: "Hunger and fear is something that every independent oil-and-gas person knows—and something that no major oil-and-gas person has ever felt in their career."

Classic Westerns are set in the wide open: on the lawless frontier, in a remote one-saloon town far from the civilizing forces of courtrooms, professional hierarchies, church committees, and social engagements. It is up to the men—and it is almost exclusively men—of the frontier to fight evil and restore order, per their own codes of justice. They exist in a world not yet bounded by rationality and law.

There's something about this world that seems like it could go on forever, even though its end is baked into its very nature. A society of dusty bachelors and gun fights can't perpetuate itself ad infinitum. Joan Didion writes of John Wayne, after seeing him sick with cancer: "When [he] rode through my childhood, and perhaps through yours, he determined forever the shape of certain of our dreams. It did not seem possible that such a man could fall ill, could carry within him that most inexplicable and ungovernable of diseases."

The film critic Roger Ebert, in his review of *The Cowboys*, points out that Wayne had died once or twice before in early films. But since he'd become a legend, his movies all ended with him alive.

Until *The Cowboys*, that is. Ebert writes: "When John Wayne finally dies, it is because *The Cowboys* has violated a Western convention. It is a sacred belief of the genre that good guys never miss, and bad guys never hit."

Just as these days in America we take electricity for granted, the small miracle of flicking a switch to turn on the light, we take the rules we were raised on as given. We signed up—or were signed up— for empire and endless growth, for a world in which better is always within reach, in which faster and cheaper unequivocally mean more for us, more bounty and more luxury and more choices. In which more is always good.

Many Westerns end not with a giant shootout, but with something more quotidian. The fight between good and evil is often unresolved. The story simply ends. The metaphor breaks down. In the case of Westerns, the sun sets and the railroads come in, twilighting the need for Wild West justice. A new metaphoric era sinks in.

ALLEN, HOWEVER, CONTINUES TO SEE POSSIBILITIES. OF THE PERMian Basin, he says, "The Permian is best viewed as a near infinite resource—we will never produce the last drop of economic oil from the Basin."

And of course, he has the numbers to justify it. Trillions of barrels. A two-thousand-year supply. Up to $100 trillion, a sizable chunk of the current US GDP. The world may be mapped, but he remains an optimist. Allen looks around and sees room for growth.

At the same time, he is not one to forget history. Since stepping back from active management at Enverus, he's taken up other interests, such as producing films. For his latest movie, *Death in Texas*, he insisted on top-notch actors, from Lara Flynn Boyle to the emerging Ronnie Blevins ("I want to make him a star," Allen says). But one name is particularly familiar to Allen. Bruce Dern, who played "Long Hair" Watts in the original *Cowboys* movie, is again cast as the bad guy in *Death in Texas*. Allen hasn't forgotten that Dern was the guy who killed John Wayne.

Death in Texas is an unexpected movie, full of heart-wrenching twists and far-fetched turns. It was largely panned by critics but adored by a subset of moviegoers. ("The older you are, the more likely you are to like it. Men hated it. Women loved it. I don't know why," quips Allen.) In addition to the usual tropes of guns and fast cars and drugs and desert landscapes, there's a dying mother and her devoted son.

The movie premieres in June 2021, in the same El Paso theater where Allen first saw *The Cowboys*. The movie starts out with the story of an ex-convict who gets tangled up with a Mexican drug cartel, in an attempt to buy his dying mother a new liver on the black market. A roller coaster of improbable events unfolds over the course of two hours. But the final twist is reserved for the last two minutes. In a scene that's hardly necessary to hold the rest of the plot together, Bruce Dern's character is shot dead. The antihero collapses, as does the curtain, and the good guys continue their ride off into the sunset.

When Allen cast Dern, he didn't give the actor the full script, nor did Allen tell him about this scene. He didn't mention Dern would die. No, the actor was given his final lines at the last minute, thus experiencing his demise as something sudden and from the blue. Just as thirteen-year-old Allen Gilmer, in a movie theater in El Paso, Texas, had experienced John Wayne's death.

The guy who killed John Wayne is dead. This is how revenge

circles round and a metaphoric era ends: fully set up, yet seemingly out of the blue.

The film credits roll. And from Paradise, Texas, TLKF croons a few sad lyrics: "This bottle in my hand, well it used to be full. Of them empty promises, to take away my blues."

5

FALSE GODS

CALIFORNIA, THEY SAY, IS WHERE THE HIGHWAY ENDS AND DREAMS
come home to roost. When they say these things, their eyes ignite:
startup riches, infinity pools, the Hollywood hills. The last thing on
their minds, of course, is the town of Stockton.

Drive east from San Francisco and, if traffic cooperates, you'll be
there in an hour and a half or two, over the long span of slate-colored
bay, past the hulking loaders at Oakland's port, skirting rich suburbs
and sweltering orchards and the government labs in Livermore, the
military depot in Tracy, all the way to where brackish bay waters
meet the San Joaquin River, where the east-west highways connect
with Interstate 5, in a tangled web of introductions that ultimately
pitches you either north toward Seattle or south to LA.

Or you might decide to stay in Stockton, spend the night. There's
a slew of motels along the interstate: La Quinta, Days Inn, Motel 6.
Breakfast at Denny's or IHOP. Stockton once had its place in the

limelight as a booming gold-rush supply point. In 2012, the city filed for bankruptcy, the largest US city until then to do so (Detroit soon bested it in 2013). First light reveals a town that's neither particularly rich nor desperately poor, hitched taut between cosmopolitan San Francisco on one side and the agricultural central valley on the other, in the middle, indistinct, suburban, and a little sad.

This isn't how the story was supposed to go. Optimization was supposed to be the recipe for a more perfect society. When John Stuart Mill aimed for the greater good, when Allen Gilmer struck out to map new pockets of oil, when Stan Ulam harnessed a supercomputer to tally possibilities: it was in service of doing more, and better, with less. Greater efficiency was meant to be an equilibrating force. We weren't supposed to have big winners and even bigger losers. We weren't supposed to have a whole sprawl of suburbs stuck in the declining middle.

In the previous chapter, we saw how overwrought optimizations can suddenly fail, and the breakdown of optimization as the default way of seeing the world can come about equally fast. What we face now is a disconnect between the continued promises of efficiency, the idea that we can optimize into perpetuity, and the reality all around: the imperfect world, the overbooked schedules, the delayed flights, the institutions in decline. And we confront the question: How can we square what optimization promised with what it's delivered?

Sam Altman has the answer. In his midthirties, with the wiry, frenetic look of a college student, he's a young man with many answers.

Sam's biography reads like a leaderboard of Silicon Valley tropes and accolades: an entrepreneur, upper-middle-class upbringing, prep school, Stanford Computer Science student, Stanford Computer Science dropout, where dropping out is one of the Valley's top status symbols. In 2015, Sam was named a *Forbes* magazine top investor under age thirty. (That anyone bothers to make a list of investors in their teens and twenties says as much about Silicon Valley as about the nominees. Tech thrives on stories of overnight riches and the mythos of the boy genius.)

Sam is the CEO and cofounder, along with electric-car-and-rocket-ship-magnate Elon Musk, of OpenAI, a company whose mission is "to ensure that artificial general intelligence benefits all of humanity." He is the former president of the Valley's top startup incubator, Y Combinator, was interim CEO of Reddit, and is currently chairman of the board of two nuclear-energy companies, Helion and Okto. His latest venture, Worldcoin, aims to scan people's eyeballs in exchange for cryptocurrency. As of 2022, the company had raised $125 million of funding from Silicon Valley investors.

But Sam doesn't rest on, or even mention, his laurels. In conversation, he is smart, curious, and kind, and you can easily tell, through his veneer of demure agreeableness, that he's driven as hell. By way of introduction to what he's passionate about, Sam describes how he used a spreadsheet to determine the seven or so domains in which he could make the greatest impact, based on weighing factors such as his own skills and resources against the world's needs. Sam readily admits he can't read emotions well, treats most conversations as logic puzzles, and not only wants to save the world but believes the world's salvation is well within reach.

A 2016 profile in *The New Yorker* sums up Sam like this: "His great weakness is his utter lack of interest in ineffective people."

Sam has, however, taken an interest in Stockton, California.

Stockton is the site of one of the most publicized experiments in Universal Basic Income (UBI), a policy proposal that grants recipients a fixed stipend, with no qualifications and no strings attached. The promise of UBI is to give cash to those who need it most and to minimize the red tape and special interests that can muck up more complex redistribution schemes. On Sam's spreadsheet of areas where he'd have impact, UBI made the cut, and he dedicated funding for a group of analysts to study its effects in six cities around the country. While he's not directly involved in Stockton, he's watching closely. The Stockton Economic Empowerment Demonstration was initially championed by another tech wunderkind, Facebook cofounder Chris Hughes. The project gave 125 families $500 per month for twenty-four months. A slew of metrics was collected in order to establish a causal relationship between the money and better outcomes.

UBI is nothing new. The concept of a guaranteed stipend has been suggested by leaders from Napoleon to Martin Luther King Jr. The contemporary American conception of UBI, however, has been around just a handful of years, marrying a utilitarian notion of societal perfectibility with a modern-day faith in technology and experimental economics.

Indeed, economists were among the first to suggest the idea of a fixed stipend, first in the context of the developing world and now in America. Esther Duflo, a creative star in the field and Nobel Prize winner, is known for her experiments with microloans in poorer nations. She's also unromantic about her discipline, embracing the concept of "economist as plumber." Duflo argues that the purpose of economics is not grand theories so much as on-the-ground empiricism. Following her lead, the contemporary argument for UBI owes less to a framework of virtue and charity and much more to the cold language of an econ textbook. Its benefits are described in terms of optimizing resources, reducing inequality, and thereby maximizing societal payoff.

The UBI experiments under way in several cities, a handful of them funded by Sam's organization, have data-collection methods primed for a top-tier academic publication. Like any good empiricist, Sam spells out his own research questions to me, and the data he's collecting to test and analyze those hypotheses.

Several thousand miles from Sam's Bay Area office, a different kind of program is in the works. When we speak by phone, Aisha Nyandoro bucks a little at my naive characterization of her work as UBI. "We don't call it universal basic income," she says. "We call it guaranteed income. It's targeted. Invested intentionally in those discriminated against." Aisha is the powerhouse founder of the Magnolia Mother's Trust, a program that gives a monthly stipend to single Black mothers in Jackson, Mississippi. The project grew out of her seeing the welfare system fail miserably for the very people it purported to help. "The social safety net is designed to keep families from rising up. Keep them teetering on edge. It's punitive paternalism. The 'safety net' that strangles."

Bureaucracy is dehumanizing, Aisha says, because it asks a person to "prove you're enough" to receive even the most basic of assistance. Magnolia Mother's Trust is unique in that it is targeted at a specific population. Aisha reels off facts. The majority of low-income women in Jackson are also mothers. In the state of Mississippi, one in four children live in poverty, and women of color earn 61 percent of what white men make. Those inequalities affect the community as a whole. In 2021, the trust gave $1,000 per month to one hundred women. While she's happy her program is gaining exposure as more people pay attention to UBI, Aisha doesn't mince words. "I have to be very explicit in naming race as an issue," she says.

Aisha's goal is to grow the program and provide cash, without qualifications, to more mothers in Jackson. Magnolia Mother's Trust was started around the same time as the Stockton project, and the nomenclature of guaranteed income has gained traction. One mother in the program writes in an article in *Ms.* magazine, "Now everyone is talking about guaranteed income, and it started here in Jackson." Whether or not it all traces back to Jackson, whether the money is guaranteed and targeted or more broadly distributed, what's undeniable is that everyone seems to be talking about UBI.

Influential figures, primarily in tech and politics, have piled on to the idea. Jack Dorsey, the billionaire founder of Twitter, with his droopy meditation eyes and guru beard, wants in. In 2020, he donated $15 million to experimental efforts in thirty US cities.

And perhaps the loudest bullhorn for the idea has been wielded by Andrew Yang, another product of Silicon Valley and a 2020 US presidential candidate. Yang is an earnest guy, unabashedly dorky. Numbers drive his straight-talking policy. Blue baseball caps for his campaign are emblazoned with one short word: MATH.

UBI's proponents see the potential to simplify the currently convoluted American welfare system, to equilibrate an uneven playing field. By decoupling basic income from employment, it could free some people up to pursue work that is meaningful.

And yet the concept, despite its many proponents, has managed to draw ire from both ends of the political spectrum. Critics on the right see UBI as an extension of the welfare state, as further interference into free markets. Left-leaning critics bemoan its "inefficient" distribution of resources: Why should high earners get as much as those below the poverty line? Why should struggling individuals get only just enough to keep them, and the capitalist system, afloat?

Detractors on both left and right default to the same language in their critiques: that of efficiency and maximizing resources. Indeed, the language of UBI's critics is all too similar to the language of its proponents, with its randomized control trials and its view of society as a closed economic system. In the face of a disconnect between what optimization promised and what it delivered, the proposed solution involves more optimizing.

Why is this? What if we were to evaluate something like UBI outside the language of efficiency? We might ask a few questions differently. What if we relaxed the suggestion that dollars can be transformed by some or another equation into individual or societal utility? What if we went further than that and relaxed the suggestion of measuring at all, as a means of determining the "best" policy? What if we put down our calculators for a moment and let go of the idea that politics is meant to engineer an optimal society in the first place? Would total anarchy ensue?

Such questions are difficult to ask because they don't sound like they're getting us anywhere. It's much easier, and more common, to tackle the problem head-on. Electric-vehicle networks such as Tesla's, billed as an alternative to the centralized oil economy, seek to optimize where charging stations are placed, how batteries are created, how software updates are sent out—and by extension, how environmental outcomes take shape. Vitamins fill the place of nutrients leached out of foods by agriculture's maximization of yields; these vitamins promise to optimize health. Vertical urban farming also purports to solve the problems of industrial agriculture, by introducing new optimizations in how light and fertilizers are delivered to greenhouse plants, run on technology platforms developed by giants such as SAP. A breathless *Forbes* article explains that the result of hydroponics is that "more people can be fed, less precious natural resources are used, and the produce is healthier and more flavorful." The article nods only briefly to downsides, such as high energy, labor, and transportation costs. It doesn't mention that many grains don't lend themselves easily to indoor farming, nor the limitations of synthetic fertilizers in place of natural regeneration of soil.

In working to counteract the shortcomings of optimization, have we only embedded ourselves deeper? For all the talk of decentralized digital currencies and local-maker economies, are we in fact more connected and centralized than ever? And less free, insofar as we're tied into platforms such as Amazon and Airbnb and Etsy? Does our lack of freedom run deeper still, by dint of the fact that fewer and fewer of us know exactly what the algorithms driving these technologies do, as more and more of us depend on them? Do these attempts to deoptimize in fact entrench the idea of optimization further?

A 1952 novel by Kurt Vonnegut highlights the temptation, and also the threat, of deoptimizing. *Player Piano* describes a mechanized society in which the need for human labor has mostly been eliminated. The remaining workers are those engineers and managers whose purpose is to keep the machines online. The core drama takes place at a factory hub called Ilium Works, where "Efficiency, Economy, and Quality" reign supreme. The book is prescient in anticipating some of our current angst—and powerlessness—about optimization's reach.

Paul Proteus is the thirty-five-year-old factory manager of the Ilium Works. His father served in the same capacity, and like him, Paul is one day expected to take over as leader of the National Manufacturing Council. Each role at Ilium is identified by a number, such as R-127 or EC-002. Paul's job is to oversee the machines.

At the time of the book's publication, Vonnegut was a young author disillusioned by his experiences in World War II and disheartened as an engineering manager at General Electric. Ilium Works is a not-so-thinly-veiled version of GE. As the novel wears on, Paul tries to free himself, to protest that "the main business of humanity is to do a good job of being human beings . . . not to serve as appendages to machines, institutions, and systems." He seeks out the elusive Ghost Shirt Society with its conspiracies to break automation, he attempts to restore an old homestead with his wife. He tries, in other words, to organize a way out of the mechanized world.

His attempts prove to be in vain. Paul fails and ends up mired in dissatisfaction. The machines take over, riots ensue, everything is destroyed. And yet, humans' love of mechanization runs deep: once the machines are destroyed, the janitors and technicians—a class on the fringes of society—quickly scramble to build things up again. *Player Piano* depicts the outcome of optimization as societal collapse and the collapse of meaning, followed by the flimsy rebuilding of the automated world we know.

SEVENTY YEARS OUT FROM PAUL PROTEUS'S ILIUM WORKS, AT THE real GE, change is afoot. Jeffrey Immelt, the company's twenty-first-century CEO, wants to simplify. Immelt is the handpicked successor of Jack Welch, the former CEO known for his often-brutal firing practices and emphasis on Six Sigma measures of productivity. Welch believed in "relentlessly driving [a vision] to completion." Now, under Immelt, efficiency is out and simplification is in.

As Immelt describes it, simplification is all about "reallocating resources to fund more growth and identify and solve customers' problems better." New times call for new methods. Whereas operational excellence and cost consciousness defined the globalizing, systems-integrating 1990s, the CEO of the new millennium sees his mandate as achieving scale through innovation instead.

The results have been mixed. In January 2022, the *Washington Post* declared simplification a "work in progress," alluding both to its lack of completion and to its focus on the "journey" rather than any destination. GE's most recent campaigns are geared, anew, at energy efficiency. Yet the company's turn toward simplicity was something more than a revamping of the wording. It asks: How would one de-optimize, optimally?

This question is at the root of the search for effective UBI. Long before Sam Altman and Andrew Yang, a group of economists and management theorists were codifying what made for an optimal process, whether of a factory line or a business decision.

Herb Simon was among them. A child of the Midwest, Simon grew up in Milwaukee, studied at the University of Chicago, and later became a professor at Carnegie Mellon. Simon's work touched on a number of fields—from industrial economics to mathematics to psychology—and his work influenced how decision-making is done in businesses, how economics is taught, and even how we think about artificial intelligence.

The research for which he is most remembered, however, is on probability and choice. Simon wanted to know how to most efficiently arrive at the best answer, for any decision process with dynamic information.

In mathematics, a false optima represents an illusory waypoint, something that looks like a summit, only to reveal an even higher peak beyond. These high, but not highest, points give the hope of having solved the equation in the best way and may have many of their own virtues—a lovely view, a mountain lake—but, in some sense, they keep us from the top.

This happens not just in landscapes of three dimensions but in equations of thousands of variables. An area of research in optimization focuses on how to search appropriately, to find the global maxima and not merely the local ones. Mathematicians have aimed to answer this question rigorously, just as Sam aims to answer the question of optimal redistribution of wealth. A class of techniques called hill climbing has an algorithm choose an arbitrary point on the landscape of possible solutions to an optimization problem, then looks around at neighboring solutions. Do any of these get you a bit higher on the hill, a little closer to the summit?

An issue with this kind of algorithm is that, for certain kinds of optimizations, it will get stuck at local peaks. You might be standing on a tiny hillock in Denver, with the foothills in the distance and the snowcapped Rockies looming just beyond, and the algorithm would insist your few steps up the hill were the best you could do.

How might you tweak your algorithm to get around this problem? One option is to run it over and over again, with slightly different parameters each time. Instead of a single starting point for your hike, you might pick a bunch of trailheads to test out. There's then the question of how to choose these starting points. Do you do a sweep of the whole breadth of the landscape first and then narrow in on those that lead to peaks? Do you crawl methodically from one edge of the map to the other, summiting every single hill to tabulate its height, before moving on to the next? Or do you sample random areas on the map, assessing which ones contain at least some high points, in order to focus on those regions to sample further?

And how do you know when to stop?

Herb Simon had a lot to say about local optima and the cost of finding global ones. Of this quest for efficient ways of decision-making, Simon writes that "the human being striving for rationality and restricted within the limits of his knowledge has developed some working procedures that partially overcome these difficulties." He goes on to mention the assumptions of any such technique: a closed system, a limited number of variables, and a "limited range of consequences." That is, decision-making is at its finest when reduced to a simple problem with known parameters.

A closed system. Isolated from the rest of the world. A limited number of variables. A limited range of consequences. These, indeed, are the criteria for most optimizations. A well-defined objective function, parameters, and constraints. Simon aims to apply these optimizations to human behavior. He coins the term "satisficing" to describe a kind of decision-making under uncertainty, in which global maxima are difficult to find, so local ones are sought instead. "Satisficing," a portmanteau of satisfying and sufficing, describes outcomes that are not optimal but "good enough" given the circumstances.

How do you know when you're satisficed? How much optimization is "enough"?

An algorithmic thought experiment, proposed around the same time, tries to answer this question mathematically. The optimal stopping problem, known casually as the secretary problem, asks the question, If you have a line of candidates to interview for a job, how do you know when to stop interviewing and make the hire?

In this line of individuals, there exists a range of qualities. To simplify, the thought experiment proposes a single metric for a good secretary, as well as a cost (your time) to interview additional candidates. Let's say secretaries are judged only on their typing speed, which ranges from fifty to one hundred words per minute. This rate can easily be translated to the dollar "value" of a better or worse secretary, in terms of how fast you'll get your letters typed. And let's say it costs you a hundred dollars per hour to interview new candidates.

Given these two variables, how do you know when it's optimal to stop interviewing? In other words, how do you determine when you've found a good-enough candidate such that the cost of interviewing additional ones, with the possibility of maybe finding someone slightly better, begins to outweigh those slight potential gains?

If you stop too early, you risk being stuck with a mediocre secretary, when a few more interviews would've yielded someone quite a bit better. If you go on too long, you risk wasting your time on interviews, when someone earlier in the process would've been nearly as good as the secretary you ended up with.

This optimization turns out to have a surprisingly elegant theoretical solution. If you have a number N of candidates, ordered randomly, you should interview N/e of them, and then pick the next best person you find after that. In this formula, e is the base of the natural logarithm, a constant equal to approximately 2.7. So if you have one hundred candidates in your pool, interview the first 100/2.7, or thirty-seven of them, and pick the next best one after that.

The optimal suboptimum.

Yet with the stopping problem, we reach a similar cul-de-sac that we saw with UBI. Each distills things down to a single metric—be it typing speed or cash. Each lays out the objective function: interview-cost minimization for the secretary problem, an optimal distribution of wealth to bring about positive societal gains in the case of UBI. While the stopping problem is a what's often called a toy model, a demonstrative example, UBI projects make bolder claims. Instead of the questions of fairness, poverty, and duty to members of one's community being considered as the multifaceted questions they are, they're often reduced to an equation to solve, a set of dollars to distribute.

If postwar economists such as Simon and scientists such as Stan Ulam created a vernacular of atomization and automation, Silicon

Valley in turn glammed up this language. As the staid concept of efficiency gave way to innovation, as everything from linguini to lingerie was gamified, the tech giants could finally say, with a straight face, that they were indeed saving the world.

This romantic take on optimization's purpose extends to much of what Silicon Valley touches, up to and including actual romance. Technologist Amy Webb's reaction to the chaos of online dating was to create a spreadsheet for love, including a list of seventy-two traits her ideal husband would possess. It was specified down to the level of taste in Broadway musicals. *Evita* was approved, but not *Cats*. She then used this list to optimize her search for a mate. Dating consultant Logan Ury applies the optimal stopping problem to finding a spouse, suggesting that if you plan to use your twenties and thirties to search for a mate, the ideal age at which to "stop" is twenty-seven, or 37 percent of the way through those two decades. At this point, she advises using one's best ex-partner as a benchmark, to settle down with the next mate surpassing that threshold.

As sociologist Eva Illouz traces in her book *The End of Love*, the modern world "encourages women and men to think of their selves as a bundle of emotional attributes to be maximized," with romantic relationships serving as the "receptacle of emotional commodities." Happiness theorist Alain de Botton claims that in the modern world, marriage has become a psychological institution, meant to optimize the growth of both parties, rather than a romantic or wealth- and lineage-preserving institution of the past. The concept of "emotional labor," to indicate the unpaid and often unacknowledged work of caregiving and consideration, has swept into the past decade's vernacular. While it captures an important idea, it's couched in the very economic language it seeks to criticize. Must even caregiving, even emotion, be reduced to the language of labor, affixed with a price tag, rather than treated as a separate and sacred thing?

As in love, so in business. An article from *The Economist* offers tips on how to remain productive in a pandemic. Tristan Harris, a former Google employee, tweets out ways to disconnect in optimal ways, with little awareness of the irony of using the very tech he promotes disconnecting from. If you need more help, there's a "solution," in the form of an app you can download, to help you stop using those darned apps so much. The very point of disconnecting is to become more productive when you're back online. At least Sam Altman is consistent in his veneration of technological solutions.

At a dinner in Austin, Texas, a Google employee learns I'm writing about optimization. "Oh, that's so exciting," she proclaims. "My job function is Optimizing Feminism at Scale." To this day, I have no idea what that mess of words actually means, beyond its revelation of tech's obsession with optimization and so-called scale, and its blundering confusion on questions of equity and gender.

Embedded in this absurdity is an earnest anxiety. We feel things are out of control, unequal, not working right. Optimization has given us the language to solve things. So perhaps it's only natural to think we can use the same language to solve those pieces of optimization that aren't quite working—to double down.

This doubling down is evident, in spades, on the issue of climate change. It's "the most important issue we face," according to UN Secretary-General António Guterres in 2018. More than two-thirds of millennials and Gen Z respondents named climate as the top priority in a 2021 Pew Research Center survey. And attention is mounting. In 2018, 74 percent of Americans said that extreme weather events over the past five years had focused their awareness of climate change as a major concern.

Why has climate has become front and center in this moment? After all, human-driven environmental impacts are nothing new. Nor is our awareness of those impacts, which have existed for perhaps as long as we've seen ourselves as separate from the natural world. The industrial revolution saw complaints about noise, pollution, the fraying of the family due to factory-driven schedules. Early advertisements for automobiles promised to solve the "horse-manure crisis" of the late nineteenth century. The so-called hockey-stick graph depicting explosive growth was popularized in the 1970s to warn about human population growth and co-opted in the 1990s as reports on global warming by the Intergovernmental Panel on Climate Change hit the mainstream.

For many decades, climate change remained a niche issue, grafted onto broader environmental and conservation-related concerns. President Jimmy Carter convened a panel and made a few soft warnings in the late 1970s. In 1988, NASA scientist James Hanson testified to the US Senate that "global warming has reached a level such that we can ascribe with a high degree of confidence a cause-and-effect relationship between the greenhouse effect and observed warming." Formerly sleepy ears in Washington perked up. The *Los Angeles Times* predicted, "Global Warming Is Expected to Be the Hot Issue of the 1990s." Yet it took until 1997 for a limp Kyoto Protocol to come together and until 2005 for it to take effect. Up until about 2018, Gallup polls showed less than 50 percent of Americans believing that climate change would pose a "serious threat" within their lifetimes. But by 2020, climate was at the fore, and discussion took on an increasingly apocalyptic tenor. Teen activist Greta Thunberg warned that "our house is on fire," alongside biting laments directed at the powers that be: "You have stolen my dreams and my childhood."

What changed? Why did corporate, political, and popular attention suddenly congeal on an issue that for decades had been relegated to the back burner, the province of academic research scientists and NGOs? Some suggest the matter didn't gain traction until corporate America could link climate to business outcomes; others point to lobbyists and policy makers as instrumental in increasing public awareness. Still others claim that the rise in extreme weather events such as wildfires and hurricanes woke people up. Yet another thread argues that over the past forty years, data and modeling have gained the precision to show incontrovertibly the effects of humans on the planet.

It's not unlikely that all these answers have merit, and indeed they're interrelated. In the past half century, our ability to collect and process data has improved tremendously, and with it has come the urge to measure and solve. Part of the renewed focus on climate is typically American-style evangelism, part of it is the kind of solutionism we see all over Silicon Valley. If it can be measured, it must be solved. When late nineteenth-century city dwellers complained of horse manure and inhumane factory conditions, these were largely local and aesthetic plaints. And the solutions, like the problems, tended to be aesthetic. By the mid-twentieth century, we'd invented quieter homes and automobiles, food that promised to be both nutritious and easy to prepare, and laws that offered reprieve from the risks of mechanized work.

Our solutions to climate change, like the research itself, have hewn closely to the tools at hand. As climate models got more complex, abetted by computational power, so did the proposed remedies to what the models predicted. Carbon taxes and elaborate cap-and-trade schemes promised to undo what humans had done to the environment. There were carbon offsets to purchase against air travel, which, like the 1990s breakfast cereal produced by members of the Grateful Dead, pledged to save Amazonian rain forests. More recent ideas have ranged from robots and sensors for home-energy savings to a mishmash of renewable energy and other incentives packaged as the Green New Deal.

In the past fifty years, environmentalism has transformed, from something local and aesthetic into a global phenomenon involving massive data sets, abstract models, corporate incentives, and treaties among nations around the world. Our urge to solve has grown in tandem with our inability to directly influence solutions. Sure, we can become vegan and bicycle to work, but these actions are drops in the bucket compared to an individual's ability to influence more local changes, from horse manure to acid rain to asbestos.

This combination of abstraction and solutionism seems primed to keep its audience hooked, without solving much of anything. Just as the most successful diet programs are those that offer paltry but promising results, just as a bank's "best" credit-card customers are those who reliably pay the minimum on time but no more—racking up profitable interest payments without incurring costly collection fees—the most successful issue du jour retains its place with a combination of urgency and impotence. We're powerless to actually "solve" climate because all of us are disconnected from the larger mechanisms of the model. We've traded slack, place, and scale for loftier solutions.

This separation is reflected in how we conduct the research itself. In climate modeling, scientists often rely on submodels called emulators. An emulator approximates a more complicated process, such as a tropical storm, by representing it as a set of more tractable equations and probabilities. These models are immensely useful for processes such as those found in climate and weather that don't lend themselves to a discoverable set of rules or a simple Monte Carlo simulation.

Emulators are marvelous tools, but they fail us in specific instances. They also encourage us to forget the more complicated phenomena that lie behind the substitution. In the same way, our transformation of climate problems into equations beyond our reach shifts our focus from tangible mechanisms to modeled outcomes. Our solutions shift from the practical and concrete to the abstract and apocalyptic.

The current apocalyptic yet individually unsolvable framing of climate also reveals how hesitant we are to tackle, or even raise, any moral and aesthetic objection to optimization. Our solutions are global and economic in nature. They're also, often, less effective. Decades of carbon offsets have done little to preserve the Amazonian rain forests and much to line the pockets of the corporations that endorse them. Robots and vertical farms may use water and nitrogen and human sweat more efficiently to make plants grow in precise settings, but they fail to restore the lost connection between farmer and eater and soil, perpetuating the ongoing miscarriage of agriculture.

The alternative is obvious but difficult. It involves a return to first principles and human-scale efforts, in lieu of abstract solutionism. Unfortunately, there's little glory, and much less money, in restoring scale.

ELON MUSK, THE ELECTRIC-VEHICLE MAGNATE, PROCLAIMS TO WERner Herzog, in the filmmaker's documentary *Lo and Behold, Reveries of the Connected World*: "The biggest risk of AI is not that it will develop a will of its own"—Musk almost pauses, a moment of awkwardness rather than dramatic impulse, before continuing—"but that it will follow the optimization function of its creators."

What Musk suggests is that the algorithms have less power than we would suppose. They're unlikely, as in the Hollywood films, to take over and unseat humans. More likely, per Musk, is that human flaws and ill intentions are programmed into the machines. The machines have a purity to them. They could indeed save the world if not bogged down by their programmers' foibles. Whereas it is we humans who are imperfect and underoptimized.

This wish to remove our biases and program things perfectly pervades the techno-utopian sensibility, down to the absurdist (if earnest) viewpoint of economists affiliated with the UK-based Adam Smith Institute, who argue that the moon should be privatized, with the sale's proceeds used to alleviate poverty on earth. When Elon Musk became the richest person in the world in 2021, he pledged, via tweet, $6 billion if the UN World Food Programme could explain, point by point, how to end world hunger. To some, it was an emperor-has-no-clothes moment for NGOs that spend more time drafting white papers than making good on their missions. To others, it was a tone-deaf troll from a man who could easily afford to feed humanity. And yet Musk and the United Nations both pointed to the same problem: neither massive riches nor massive planning is enough to meet this most basic necessity. And both leaned on the the same language to say so, one of spreadsheets and numbers and trade-offs.

When Sam cofounded OpenAI with Musk, he saw artificial intelligence not just as a set of cool tools but as another way to save the world. AI technology was advancing rapidly and would only continue to do so. It would soon create sufficient wealth for every adult to receive $13,500 annually, as Sam suggested in an online post that was picked up by major news outlets. He later tells me this number was taken out of context. He didn't mean it exactly. But the idea holds. Automation will give humans far more leisure time for creative endeavors, directing their productive energies toward new invention.

In the decades since Ulam, computer algorithms have built up layers of Monte Carlo–type simulations, layers of linear programming to find maxima, layers of abstraction. The difference is more than one of degree. There's been a shift, too, in how the modeling is done. Earlier models were not unlike how I'd first imagined the Amazon bulldozer back in Kentucky, following the existing shape of the field, emphasizing an existing outline with a deeper rut. When we built cities and farms, we looked first at what the site suggested, which way the wind blew, how the water would flow.

These days, we start off with what we imagine is a blank slate. We throw in piles of data and let the computer code spit out the "best." What's the result? AI-generated text that emulates perfectly some author of the past, art that copies and enhances at the edges. It's immediate, it's comprehensive, but it's fake. It's not just fake in the way that rows of painters replicating Monet water lilies on canvases in Saigon are fake; it's fake in a way that robs us of our own creativity. It crowds out the older, harder ways of creating. It's highly dependent on its inputs and therefore lacks the regenerative, going-outside-the-system force that propels most human art.

Sociologist Elizabeth Popp Berman, author of *Thinking Like an Economist*, describes the economic-policy consensus that dominated Washington, DC, from the mid-twentieth century to today, across both Republican and Democratic administrations: "a focus on leveraging choice, competition, incentives, and the power of markets in the pursuit of outcomes that would be not just effective, but efficient." She argues that this point of view has pervaded American policy for the past half century.

And it's not just politics but every major industry in America today. Sam's own optimal thinking extends to his eyeball-scans-for-cryptocurrency company, Worldcoin, where things have started to go off the rails. The company's pitch made sense in Silicon Valley. Investors latched on to the idea of uniquely identifying every human on earth with a biometric signature. Yet elsewhere, the "orbs" used to collect data, such as eyeball scans, are being met with strong resistance, both by individuals and by governments in the mostly developing countries in which they've been rolled out. The promised cryptocurrency payments haven't been delivered.

The company claims to be trying something grander, in the vein of UBI. It wants to help distribute wealth more equally, and, as CEO Alex Blania says to BuzzFeed News, the eyeball gathering is a way to authenticate that someone is "human, unique, alive." This kind of authentication would be the first of its kind, and monumental to all sorts of internet technologies, Blania believes. There's even boiler-plate to get authorities off their backs: "We all benefit when everyone takes part in the Worldcoin project."

Buzzfeed News authors Richard Nieva and Aman Sethi conclude, morosely: "All told, there is a disconnect between the Silicon Valley executives convinced that billions of people from diverse cultures will embrace their grandiose vision and the difficulty that Orb op-erators have faced on the ground as they gather eye scans from people around the world."

Sam Altman and his ilk are the modern-day bearers of the torch of optimization, with a concomitant hesitation about being on the winning side of free-market libertarianism. This hesitation takes shape in a doleful acknowledgment that optimization, too, may have its limits. Sam's take, for example, on the effective altruism movement is less than sanguine. Effective altruism is a loose set of philosophical principles practiced by a small group of primarily tech workers and intellectuals. It is, according to the EA organization's website, "focused on maximizing the good you can do," something achieved through careful consideration of how to apply one's time and money toward the best possible work and philanthropy. It's dizzying to read so many "bests" and "maximals" in one go.

Sam's held back on getting into the movement. He's a realist, with enough savvy to know that not everyone sees everything as a calcula-tion. Some things, he believes, can't be computed as universals, and some things are simply a matter of personal preference. Personal preference that can, of course, be fleshed out by spreadsheet.

OPTIMIZATION'S ILLUSIONS RUN DEEP. IN THE PAST TWENTY YEARS, the nature of mathematical modeling has changed tremendously. Computing power has grown several orders of magnitude. Not just the outcomes but the parameters of the model itself are optimized. All of this has served to divorce the practitioner of optimization even further from an understanding of the material world beneath the math.

The further our models get from reality, the more optimization's promises fall flat. And the more, it seems, we continue to pile on fixes: more planning committees, measurements, surveys, clinical trials, and experiments. Satisficing, or deoptimizing optimally, feels like a fix. If we can only determine how much to optimize, when to stop at the perfect local optima, we'll have nixed the issues with optimization while maintaining its many virtues.

All of this encourages us to keep thinking in optimal terms, even when a completely different way of thinking may be required. Under the most narrow of definitions, we may have found the optimal suboptimum. But we've also cemented our faith in optimizing, while learning little about how else to make an evaluation in a dynamic, changing world.

There's another problem with these deoptimizations. All too often they're co-opted to serve those who control the optimizations in the first place. Take the mindfulness program at Google, complete with a Zen monk turned meditation adviser. Meditation is pitched as a perk and used to attract talent and increase productivity. These ends serve Google as much as, if not more than, they serve the individual employee. Likewise, so-called circuit breakers for overoptimized financial markets typically help the biggest players, long before the little guy knows what's happening. In 2021 and 2022, the Federal Reserve claimed to be engineering a "soft landing" for the US economy, by raising interest rates until inflation hit a target rate. According to Nouriel Roubini, an economist known for his pessimistic outlook, the Fed's plan is "mission impossible." There's no way to fine-tune such a complex system. It is also, no doubt, an example of how far the drive to control and engineer has come.

Why do we embrace these engineered solutions so readily? Part of it is that we're swimming in the vernacular of efficiency. And part is that we want to feel as if we're participating in the way out. By offering a preprogrammed escape route, these deoptimizations give us the illusion of forward progress.

Sam Altman's team has run its nationwide experiments and gathered "tons of data." And yet I can't help but wonder how the distribution of stipends fits in with the gospel of growth, as summarized by Sam in an online essay: "Democracy only works in a growing economy. Without a return to economic growth, the democratic experiment will fail."

How do you solve the problems wrought by optimization, by a world singularly focused on dollars and growth?

Silicon Valley's answer is simple: not-enough-dollars is the problem. The solution is to give people more dollars, to redistribute dollars more optimally. The question of how many dollars can be answered experimentally. We can test the impact of $600 versus $500. Will

$600 decrease monthly fluctuations in spending by 36 percent instead of 46 percent?

For Aisha and the Magnolia Mother's Trust, the language is a little different. A mother was able to make rent. She bought a pizza dinner for her child's birthday. And yet, when it comes down to it, Sam's world controls more dollars. In Jackson, just as anywhere else, they're subject to Silicon Valley's solution.

JUST SOUTH OF SAN FRANCISCO, THE RAGING PACIFIC OCEAN LOOKS almost turquoise, shining in the sun. That same sun glints off metallic office buildings, newly poured pavement, and the swift straggle of cars winding past eucalyptus groves and golden hillsides.

Save for the sprawling tech campuses and near-identical condos and ranch homes, Silicon Valley possesses a natural beauty that's almost painful to behold. As you crest Highway 280 toward the ocean framed by rolling hills, as you continue down the coast to the cliffs of Monterrey and Big Sur, as you inhale the sun and surf and eucalyptus and oak, there's an achy gasp to believe this place can actually exist.

Maybe it becomes easier to believe, in a place of such physical splendor, that the world we inhabit is, if not perfect, perfectible. It's certainly a belief that's held dear in Silicon Valley. Perfection requires a far smaller stretch of the imagination here than it must have for James Mill laboring away in the nineteenth-century gray of England.

Indeed, the idea of a perfectible world is embraced past the point of absurdity by Silicon Valley's elite. Engineers speak of "making the world a better place," a phrase parodied by the HBO show *Silicon Valley*, but also ever present in techies' vernacular. Any talk of getting rich or getting famous is stuffed keenly under the veneer of improving humanity's lot.

The gospel of optimization is present in what Silicon Valley consumes, as well. Soylent and Bulletproof Coffee and special vitamin packs and paleo diets optimize health. Engineers search for the optimal data-storage solution. HR managers optimize the workforce. Together, it's said, we can optimize the world.

There's a gaping chasm dividing somewhere like Palo Alto, California, from much of the rest of America. The chasm is racial and structural, economic, geographical, and cultural. It's a division between the winners and losers of the past two centuries of optimization. It's seen in the disconnect between what optimization has given us and how it's failed. We have modern medicine, office jobs, credit cards, mortgages, and new cars. We optimized railroads to cross the continent and machinery to extract minerals from deep down. We made R&D more efficient. We built computers in the clouds and streamlined cheap labor abroad. And we feel more disconnected than ever.

And yet . . . it wasn't supposed to be this way. Optimization was supposed to be an equalizing force. As articulated by John Stuart Mill's colleague Jeremy Bentham, the intent was for "everyone to count for one, no one for more than one." Indeed, it was the idea of utility, and the efficiency therein, that allowed Mill to reconcile two of his own otherwise contradictory beliefs: the belief in a top-down means of planning with the laissez-faire belief in individual self-determination, in each one's quest to seek out happiness.

While the proponents of optimization continue to promote the kinds of techno-utopias parodied in *Player Piano*, critics begin to pop up. We've already met a few, such as farmer Bob, who tries to hold out on GMO, and Ed Grothus, with his antinuclear church. Often, the critique is romantic in tone, advocating a return to simpler times. The back-to-the-land movement of the 1970s brought a small resurgence in homemade cheese and collective living schemes. More modern movements, from the Cottagecore aesthetic of flowered dresses and sourdough starter to the American Redoubt separatist campaigns, seek similarly romantic escapes, with varying political ornamentations.

A number of critics have taken aim at tech companies' practices around data and privacy. Shoshana Zuboff coined the phrase "surveillance capitalism" to describe the commodification of personal data for the gain of corporations. These accounts, while accurate and disturbing, fail to identify a larger fallout of seeing through the lens of measurement and prediction. Having worked with these algorithms and all their flaws, I'm somewhat more sanguine than Zuboff is about the overreach of the technologies. While the potential for digital surveillance is treacherous, the skill of machine learning has been vastly overstated to the public: many of the models just aren't all that good, especially when it comes to cleaning and joining messy real-world data. It seems we're already growing bored of Facebook's metaverse, of expensive escapades such as OpenAI's GPT-3 and DALL-E, which leave us with little more than auto-generated news articles and generic, unattractive works of digital art.

What is far more troublesome, I think, is our belief in the superpowers of these programs. Our choice to see things this way not only reinforces the lens but also gives space to human institutions of control to confirm and take advantage of that faith and, ultimately, of us. The related problem is that even once we lose our trust in the particulars (as we have begun to, in tech and social media), the myth persists.

My first hint that tech solutionism, and optimization in general, was faltering came a decade earlier. It began to take shape on a cruise ship somewhere off the coast of Florida. We were two days into the cruise when I learned that the ship had been slowed down to half its clip. On purpose.

This was the first cruise ship I'd ever been on and, as I decided just twenty-four hours in, the last I'd ever board. We were less than a hundred miles off the coast, in the spring of 2011.

The ship felt massive, although it was actually one of the smallest around. As our taxi driver pulled into the cruise docks in Miami, he called out vessels at a distance: this one is Carnival, that one Royal Caribbean. He finally landed on ours: the runt of the litter, dwarfed by the behemoths with many times the population, metropolises of waterslides and ballrooms and arcades.

Size aside, this was no ordinary cruise. I'd been invited—my passage sponsored—by a group of young tech bros intent on starting a new conference series. They wanted it to be the next TED, except on a cruise ship or mountaintop (they'd later purchase an entire ski resort in Utah). Of course, the bros didn't call themselves bros. Nor, for that matter, did they have much in the way of actual experience in technology. But they were fantastic promoters. Aboard the ship, they said, would be all manner of glitterati: CEOs, VCs, top scientists, and big names in music, politics, and film.

I was, as they saw it, emblematic of the bright future they wanted their brand to represent. Young, female, and working fluently with data to "build a better world."

From the taxi, the scale of these ships seemed fantastical, but once in their midst a claustrophobia set in. We entered a tent, a long snaking line of beautiful people waiting to board. At the mouth of the line was the passport check; leading up to it, a cesspool of pep and charm. The organizers made their way up and down the line, shaking hands with gusto and bright eyes that looked right past you. Attendees breathlessly dropped job titles and place names like Aspen and Bali. We were told to download an app that let us share résumés instantly, by touching one phone to another. I didn't own a smartphone, so my scholarship-kid flip phone stayed tucked in my purse. The app turned out to not really work, anyway.

Above it all was a large banner that read MAKE NO SMALL PLANS.

On board, music blasted from loudspeakers. Richard Branson dropped big Life Lessons in a cavernous auditorium, as the engines warmed and the crew scurried to scrub floors and fill buffet trays. It was rumored, but never confirmed, that Branson departed early by helicopter, instead of attending the cruise in its entirety. The ship burned tens of thousands of gallons of fuel per day. It created mountains of trash. It staffed hundreds of people from around the world, working long hours at minimum wage and mostly invisible to the big names on board.

On the third day, we were given the option to go "onshore" to a cruise-line-owned cay, a sad piece of dirt in a stunning turquoise sea. The crew set out listless trays of pineapple, upon which a cloud of black flies immediately descended. Down the beach, a lifestyle guru gave an inspirational talk, upon which a cloud of eager humans descended in kind. In transit back from the cay, discussion bubbled up about the ship's trajectory. Someone in the know revealed: we're actually going half the normal speed. But that's cool, they added, to nods of agreement. It's not like we're going anywhere in particular. We're here to meet interesting people and build our networks.

Others confirmed. We were, indeed, going as slow as possible, cutting proverbial circles off the coast of Florida. This fact, once out, was not suppressed but promoted as an unadulterated marvel. We were there to meet like minds. Slowing down meant we'd be less distracted by things outside the ship, more able to focus on what mattered, maximizing the productive connections on board. We were burning more fuel to get to the same destination, which was nowhere at all.

For all the absurdity of the cruise, the scene and the people on board stand testament to a set of principles the early twenty-first century holds dear. In 2011, as Instagram was just getting started and years before the likes of Facebook and Google faced scrutiny over business practices, the tech world seemed, to many, infallible. We'd wiped away the financial sector's crisis with a quick bailout, even if the country beyond Wall Street was faltering. The smart kids got out of banking and tilted west toward a new pot of gold in Silicon Valley. Information technology was seen as an equalizing, even revolutionary, force. It was set to change the world, and it would do so by making computation and broadband connectivity as cheap as possible for as many people as possible, around the world. And to do all that, the people on top just needed to slow down.

For me, the cruise was a turning point. It crystallized the things we'd lost with optimization and how ill-equipped we were to refind them. It also deepened my disillusionment: the ones running the show, steering the ship, didn't get it. They were too busy exchanging contact information and swaying to Richard Branson's lulling platitudes. There was a smugness, a moral superiority, to the idea of artificially "going slow." Slowness, here, was an engineered shortcut, a belief that forging human community was merely a matter of getting the best people in the same place.

Worst of all, I couldn't see my own way off the ship. There seemed to be just two options: opt out completely or double down on optimization.

The tech world offered ample examples of how to double down, and as I hopped from one center of maximizing at MIT to another in Silicon Valley, I despaired that doubling down was the only thing available to me. After all, if I were to try to leave, where would I go? San Francisco was my hometown, optimization my native language.

More than that, there was an enchantment in doubling down. Part of me, not unlike Sam Altman, believed the world was ours to shape. That there was a certain beauty in organizing things mathematically. Or, as a scientist of an older generation once put it to me more modestly and more elegantly, "My work is to reduce entropy in some small corner of the universe." That is, to hem in some of the disorder. To fit that corner of knowledge into a tidy frame.

And yet I couldn't shed the disconnect between the promise and reality of the cruise. The tanks of fuel ignited to "slow down." It felt deeply unsettling to be trapped on a huge engineered ship, going nowhere.

I'll never board another cruise ship, but I have been on a few odd sailboats. Something strange happens at night in the middle of the open ocean. If the ship is small and there is no moon, the inky sea begins to blend with the black sky, erasing all marks of space and time. Sailors who've spent weeks at sea speak of distant shorelines that turn out to be illusions, just as desert wanderers see glimmering oases. On the cruise ship, by contrast, we were not an hour from shore, stuffed with delights and self-assurance, and fancied ourselves so righteous as to construct mountains everywhere.

In the twenty-first century, a funny thing has happened. Optimization's true believers have adopted some of its critics' language. Those at the pinnacle of optimizing's pyramid, such as Jeff Immelt and Marie Kondo, have latched on to the simplification zeitgeist. Sam Altman's quest to defuse inequality is not just technocratic, but highly sentimental. He believes optimization can help us escape the problems that it has wrought. With the sun glinting off the ocean and into our eyes, on this Western edge of the map, this fallacy may be hard to see. But standing in Stockton or Jackson on a gray humid summer day, it's crystal clear.

6

THE TREACHERY
OF OPTIMALS

"SHH," SAYS JASON. FOUR BISON GAVE BIRTH LAST NIGHT, AND THE whole herd has its hackles up.

We're parked a bumpy dirt mile in from the highway, looking over a small bluff at four wobbling calves. Not twelve hours into life on Earth, they're already running tentatively, watched relentlessly by the adults. The wind howls, and I try to keep my voice loud enough for Jason to hear but quiet enough not to spook the animals.

Jason is a biologist and a member of the Eastern Shoshone. Two hundred years ago, the Shoshone spanned lands from the Great Basin of Nevada across the Rocky Mountains and into the plains of Wyoming. Today, Jason lives in the Wind River Reservation, where he founded and manages the tribe's bison-restoration project.

A soft-spoken man, Jason gets fired up when he talks about buffalo. His people have long been entwined with the species. In addition to being a source of sustenance, the buffalo is an integral part of their culture. Its hides were used for shelter, its flesh for nutrition, its bones in ceremony.

Forty miles away, Lander is a town of about eight thousand people on the north side of the Wind River. It's home to offices of the US Forest Service, the FBI, the US Fish & Wildlife Service, a bronze foundry, and some bars and restaurants and hardware stores perched on the plains. Just across from the Ace Hardware on Main Street, there's a mural of three bison and a calf. It's painted in the browns and yellows and olive greens that characterize this landscape. One buffalo is seated and two stand regal, with what looks like sadness in their eyes.

Once, these plains were filled with bison. Now, a small herd that Jason tends occupies some three hundred acres off Highway 26.

The son of a scientist and outdoorsman, Jason grew up outside. But it was a trip with his father to East Africa that would set his life path in motion. In an interview with his alma mater Montana State, he recalls witnessing the wildebeest migration on the Serengeti: "The 1.5 million animals in the wildebeest migration . . . is just a large, abstract number until one witnesses their journey in person. We drove for miles through the migration for nearly two days, seeing a multitude of other species in participation. The sheer number was awe-inspiring."

After this trip, Jason realized the Americas were not without their own Serengeti. Just two centuries ago, the bison population on the North American plains was twenty times the number of the wildebeests in Africa. Jason sees the plains as they were before the encroachment of ranchers and cattle. He sees the plains of the future once again filled with bison herds—a fenceless self-sustaining eco-system, where tribal members and buffalo and other plains species live in symbiosis.

This vision entails more than just plopping more buffalo in fields, Jason tells me. It involves what he calls a decolonization of land. It means unwinding the myriad uses codified in the past two hundred years, from the diversion of waterways, to the grazing of cattle, to the irrigation of farmland, to the mining of coal and development of highways.

The American buffalo, or bison, once occupied a rangeland stretching from Alaska to the Gulf of Mexico. As late as the mid-1700s, there were sightings of the animal as far east as Florida. It's the largest and heaviest land animal in North America. Males some-times weigh in at more than two thousand pounds. The animals can jump up to six feet in the air and run up to fifty-five miles per hour.

In the late 1700s, there were an estimated seventy million bison in North America. By the late 1880s, there were just a few hundred left. Western settlers killed hundreds of thousands of buffalo in a matter of two decades. Many call it a slaughter, barbarian in its speed and scope. It was also, in some sense, an optimization: a blank slate was cleared and divvied up into something more profitable—at least when it came to dollars and in the short term.

Now, in the absence of bison, we're left with a question. Is it pos-sible to unwind that damage done to species and ecosystem, and restore both to their former state?

The previous chapter looked at one way of resolving the short-comings of optimization, by attempting to deoptimize in an optimal way. A second response to optimization is less equivocal. It says, If an optimization doesn't work, we shouldn't try to remedy or patch. Instead, we must undo.

If the first way stays wedded to the framework of optimization, the second seeks to clear the slate. The first is a response born of control and engineering. The second is a drive to retreat, to let go of the reins, to rewind to some past reference point. What the two approaches have in common is that they're more difficult than they appear.

As we'll start to see in this chapter, an optimization has directionality, an asymmetry in time. While there's a formula for optimizing, there's no clear recipe for unwinding. We know how to build the skyscraper but not how to dismantle it gracefully.

Let's say, for example, we agree that our modern system of agriculture isn't working. Too much grain is shipped abroad, too much soil depleted by giant tractors and megafarms. Let's say we decide to reverse all this, by encouraging smaller farms and more local distribution. If we're trying to envision such a system, a number of practical questions arise.

How do we decide what is the "right" size of farm? What balance should we seek between environmental impact and food cost? Who is going to pay to replace the infrastructure that's grown around massive farms churning out cheap yields—the $500,000 tractors, the transportation and food-processing systems that rely on consistency and scale?

There isn't just one metric involved but many, and some of them are quite difficult to suss out. Some are not questions of measurement at all but choices around values. How many farms and farmers do we need? Is it a world in which almost everyone farms, as was more or less the case for a few thousand years? If everyone farms, who will be a doctor or a plant biologist or a baker? If it's not everyone farming, then what is it? One farmer for every ten people or every hundred? How do we decide? Do we let the free market pick? Do we find the "optimal" diversity of farm sizes and dictate from on high?

Seen in this way, the question starts to become absurd. In practice, it's a patchwork of enterprises and policies that eventually springs up, pushing things in one direction or another, often with much variety and as many steps forward as back. In American agriculture, this has meant a bumpy road toward some smaller farms and markets, whose products remain accessible to only a fraction of the population. It has meant family operations pushing back against the seed companies on the one side and using federal interventions to their advantage on the other.

Compare that kind of messy unwinding to how an optimization develops in the first place. Less than a hundred years went by between Borlaug's research on crop breeding and the advent of giant farms. Each year farm sizes and yields marched upward with very little lag or retreat. Optimization is easy. We can optimize for revenues, for yield, even for water use or nutrition or average human life span. Deoptimization is much harder.

This asymmetry is captured by a painting in a far-off gallery in Los Angeles. A beige-toned canvas depicts a wooden pipe. The hues are not unlike those of the bison painted in Lander, but the picture is different. The surface of the pipe gleams, and the object is underscored by five French words: "Ceci n'est pas une pipe."

This is not a pipe. The image by Belgian surrealist painter René Magritte looks very much like a pipe. Yet as the words below remind us, it's not a pipe at all but a depiction. Magritte alternatively titled the painting *The Treachery of Images.*

If we saw in Silicon Valley the problems with solving optimization's downsides with more of the same, we see in Wyoming an additional complication: the way out of optimization it not as simple as the way in. This we might call the treachery of optimals. Like a work of art, an optimization is often, treacherously, mistaken for the reality it seeks to capture. The more we optimize, the more we frame things in a certain way. The more we label it a "pipe" and not just a mess of colors on a canvas, the harder it becomes to see it any other way. The better we get at optimizing locally, the worse we get at reconsidering the optimization itself.

Just as a painting is deceptive, substituting an image for the real thing, optimization deceives by suggesting its outcome is so framed because it's the best and most real way there is.

And like a painting, an optimization seeks to bound its worldview within a tidy frame or model. The inflation and unemployment projections of economists, which reflect and also shape how we feel about the future, are little more than rough mirrors to current and past parameters. A train schedule may hold our day in one way, a school schedule in another, a night shift in yet a third. None of these structures are necessarily more true than any other, and still our bodies and our habits shape around them.

THE SHOSHONE PEOPLE ONCE FLOURISHED ACROSS THE AMERICAN plains. Through a series of diseases brought by settlers and war, both against the United States and allied with the United States against other tribes, the population was fragmented, and land shrank in a matter of decades. Today, some twelve thousand people identify as Shoshone, spread out across the West.

A not dissimilar fate befell the American buffalo in the late nineteenth century. A photo from the mid-1870s shows a fifty-foot-high pile of bison skulls with two perched beneath. *The Destruction of the Bison*, by environmental historian Andrew Isenberg, quotes a firsthand account of the slaughter: "Where there were myriads of buffalo the year before, there were now myriads of carcasses. The air was foul with a sickening stench, and the vast plain which only a short twelve months before teemed with animal life, was a dead, solitary desert."

Thousands of buffalo were killed every day in the early 1870s, and in just a couple decades the population dropped from tens of millions to a little over three hundred in 1884. They were killed for their meat, for their hides, and simply because they were tied into Indigenous peoples' survival. Although never an official policy act, slaughtering buffalo was seen by many as an act of war against the tribes; President Grant considered it a "solution" to the "Indian Problem." General William T. Sherman, of Civil War infamy, went even further, writing in a letter: "I think it would be wise to invite all the sportsmen of England and America there this fall for a Grand Buffalo hunt, and make one grand sweep of them all."

Today, the population has recovered somewhat, and an estimated four hundred thousand to five hundred thousand bison live in the United States, with twenty thousand to thirty thousand of them roaming free. The vast majority are held on private land. Jason wants to see an increasing number of herds roam free of fences, as they did 150 years ago.

The steps to this vision are not trivial. First and foremost, it involves restoring what are now grazing areas for cattle to wildlife land, both in practice and as an official designation. This means agricultural resource plans, it means negotiating with ranchers who profit from their cattle's grazing the land, it means water management and money to purchase land.

And it's more than that. When Jason talks about decolonization, he isn't just talking about land ownership. He means the decolonization of a particular way of seeing land: as a resource to be mined, as a plane from which value can be extracted. He hopes to restore the buffalo not just to its habitat but to its central role in his culture. There are plans for a buffalo institute and culture school, and efforts to bring back traditional ceremonies. Jason speaks of it as returning a center of gravity to the tribe, of which 90 percent are under the age of thirty. "We're losing our elders day by day," he says.

Earlier, we encountered William Stanley Jevons's paradox around the use of natural resources. His insight was that as the price of a scarce resource, such as coal, increased, people got more inventive about mining it more cheaply. A more efficient way to mine resources in turn lowered prices. When the price goes down, infrastructure and expectations build. We add new lights and appliances to our homes. We become more reliant on those new lights and the cheap price of electricity.

So it is with optimization. We'd expect that once we've optimized things, once we've developed a system to manufacture components and assemble iPhones for the lowest price abroad, then we'd finally be happy and rest. But now that this optimization is cemented, now that the drive to optimize is cemented as a worldview, we want more.

More doesn't always work out as planned. While abundant rangeland and water diverted for irrigation and cheaper grain for feed have lowered the price of beef in the United States, they've also degraded hundreds of thousands of acres of soil and come at the cost of hundreds of thousands of head of bison, not to mention entire ecosystems and a way of human life.

ON A MAP, WYOMING IS A SIMPLE SQUARE OF A STATE, INCLINED BE-tween mountains and plains. There's the vast green wildernesses of Idaho and Montana to the northwest, the red canyonlands of Utah to the southwest, the plains of South Dakota and Nebraska to the east. Heading north from Colorado on the interstate, you soon see signs for Wyoming's first patch of civilization: the Terry Bison Ranch.

Just off the highway exit for the ranch are three enormous warehouses hawking fireworks. Billboards with the Terry Bison logo lead the way in from the frontage road. Some five miles in, the road loops back again to face the highway, and you come to what was formerly a working ranch. Now it's a gussied-up RV park, complete with touring train and horse rides.

The entrance is flanked by a replica of a Hollywood Western set. Painted colorfully on the facade of Main Street are its saloon, hotel, and barbershop. Just underneath the HOTEL sign, a walkway leads past a swinging door to an unusually abundant array of road-stop fare: etched shot glasses and T-shirts, baby onesies, beef jerky, and plastic cowboy hats with "Wyoming" emblazoned across the crown. The gift shop serves as check-in desk for the campground, and the woman behind the counter has no bad news. The band's fantastic tonight, the kitchen's open until 10:00 p.m., the rib eye not to be missed. Make sure to stay for dessert, too.

It's not until morning light that one comes to understand where this funny oasis got its name. A pink sun rises over the parked RVs and Western storefronts. Down the bend from the restaurant is a small herd of bison, a petting zoo almost. They scuff around in the dirt, and watching them is one of the activities on offer, along with the saloon and swimming pool.

I wander past the bison to a small shack of a café. It's just before 7:00 a.m., and the open sign is unlit. As I approach, the sign flicks on and a woman in an apron with tattoos and long, thick, braided hair steps outside to sweep the porch. "I'm not quite open, honey," she says. "Give me ten minutes."

At 7:00 a.m. on the dot, she scurries me inside. I order a coffee to go and an egg-and-cheese sandwich, then wander out to check on my dog, who's by now agitated by a gaggle of itinerant chickens and barking madly from the truck. A ranch hand in a baseball cap wanders by and asks, "Where are you from?" And after a glance at my license plate he answers his own question: "Wow, that's a ways!" A Wyoming state flag flaps in the wind: the outline of a bison.

When I go back inside, the café woman and I get to chatting. Growing up, she'd spent a few years working on this ranch. Then she married a man in the military, and they traveled to all kinds of places before landing back in Cheyenne. In the course of a couple years, her husband was killed in service overseas, her parents in an accident. She didn't really know where to go. So back to the ranch she came, and she's been here ever since.

This information is conveyed matter-of-factly as she pours hot coffee into a paper cup, tops it with a plastic lid, and wraps up my breakfast sandwich. It's spoken as if giving directions to the nearest gas station.

She ended up at the Terry Bison Ranch for no particular reason other than that it was a place she'd once been. It was, long ago, a place, and had now over time become a nonplace, of travelers-through and cowboy bands making a one-night stand, of mocked-up storefronts and sad-looking buffalo behind a fence. But she'd put down roots and made a *there* here.

I smile and tuck my sandwich under one arm, keys in one hand, coffee cup in the other. I'd spent so long wandering around the country that it had become easy to forget where exactly I was now. And here I found myself outside Cheyenne, still in motion, perpetually heading out, always a little behind schedule.

President Roosevelt stayed at the Terry Bison Ranch twice, in 1903 and 1910. The ranch once stretched 300,000 acres and was owned by F. E. Warren, the first territorial governor and the first US senator from Wyoming. It's now down to a still-sprawling 27,000 acres and home to about 2,500 head of buffalo, most of which steer clear of the petting-zoo-like fences and the humans who gather there to gawk.

In the early 1900s, Western settlers started to take an interest in bison, a species that had been nearly extinguished just a few decades before. Ranchers such as Warren acquired and bred herds, appreciating the animal's utility, as well as its majesty. In the decades that followed, bison ranching became an upper-crusty pursuit, distinguished from the more pedestrian cattle operations. Ted Turner, the media magnate, is one of the largest landholders in the United States and also the owner of more bison than anyone else on earth, some 45,000 head.

Turner belongs to this older school of bison ranchers, in which animals are often interbred with cattle. Although prized for their distinct meat and hides, more often than not, they're husbanded much as cows are. And for all its hats and bravado, the cattle industry isn't among America's most environmentally sustainable.

A newer generation of bison ranchers is trying to change all that. They want bison affiliated with neither the Ted Turners of the world nor environmental degradation. These ranchers also want to see more humane treatment of the herds. Matt Skogland of North Bridger Bison came into the profession with no deep pockets or connections—he grew up in suburban Chicago—but with a vision to ranch the animals sustainably. Skogland incorporates field slaughter into his ranching: he goes out with a rifle and a single bullet, fells the buffalo, then spends the next hours field-dressing and packing it out. His customers are local households near Bozeman, Montana. The ranch prides itself on regenerative agriculture practices, working to rebuild soil and support a diverse ecosystem of animals and plants.

There have also been attempts, like Jason's, to set aside land. To the northeast, in the Montana plains, the American Prairie (formerly called American Prairie Reserve) is working to connect millions of acres of public lands so that bison can roam free. Sean Gerrity, a former technology executive and Montana native, founded the organization in 1999, hoping to patch together enough land to sustain not just bison but related species as well. By 2012, the organization had managed to corral 250 head on 60,000 acres, and by 2020, 800 head on about 450,000 acres. Gerrity eventually sees 25,000 bison, and multiple packs of wolves, all occupying 500,000 acres, with an adjacent 3 million acres of Bureau of Land Management and Montana state lands also open to the herd.

Gerrity is a gregarious guy, fun to talk to, as earnest as he is smart. He believes it's an "inevitability" that bison will one day be free roaming, rather than confined to national parks, such as Yellowstone, or private ranches. He's also focused on keeping the genetics of the bison strong, and independent of cattle.

For Gerrity, this is as much about politics as it is about science. If bison are seen as a distinct species, they are more likely to be treated as such and politicians to make laws to support them. The problem is not people, he says. After all, "people coexisted with those animals for ten thousand years, but this group of people came with barbed wire, set it up, said this land is going to be surface agriculture with farming and livestock," Gerrity says.

A genetic bottleneck occurs when a species becomes too few and there's not enough diversity to keep a healthy herd. Inbreeding becomes more likely, recessive traits more pronounced, offspring less likely to be viable. Disease can more readily decimate a population. Bison need to survive harsh winters and trying conditions on the plains, and when their numbers dwindle to just a few hundred or thousand, the species is threatened.

These days, the American buffalo's odds seem much improved, their numbers have increased, and it looks like some of the species' traits for adaptability persisted through the lean times. According to

Gerrity, their odds are better for another reason, too: more attention from the public, and attention of the right kind.

Gerrity sees the importance of negotiations with cattle ranchers, grant writing and influencing policy, and acquisitions of nearby land. He is sober and practical. Coming from the world of business, Gerrity brings a distinct understanding of what ultimately drives things, too: money and land. "We're opening a high-end safari lodge like they have in Africa," he says. "We want to make it a world-class reserve like people have never seen in North America."

Like Jason, Gerrity takes the wilderness and wild fauna of the African continent as inspiration. Yet where Gerrity sees the scientific argument for conservation falling short because it's not sufficiently political, sufficiently convincing, Jason bemoans a different issue. "Ecology is a discipline that mirrors Native philosophical values but lacks the spiritual interconnectedness that is prevalent in Native ways of knowing," he says. Science, in other words, is precise and practical in telling us how to conserve a species, but ultimately hollow in answering why.

Although the American Prairie has mostly focused on existing public lands, it's bought up some private ranches, as well, from anyone willing to sell. A rancher named Conni French, interviewed by NPR, isn't happy: "I see them coming in with big money, buying up ranches and walking over the top of the people who are already here . . . For them to be successful in their goals, we can't be here, and that's not OK with us." A comment on the Ranchers Stewardship Alliance Facebook page confirms Conni's sentiment. American Prairie Reserve "will not be happy until every family ranch surrounding it is bought out," the poster writes. "It will ruin the rural communities and towns as we know them." Protest signs along area roads read "Save the Cowboy."

It may seem strange to say, but Conni French and Jason have more in common than the two might first imagine. In the same way that Jason protests the optimization of land to meld around cattle and ranching interests, Conni decries the newfound optimization of buying ranches to restore bison in her backyard.

Sean Gerrity's stance is that, in some ways, we can solve the problem with more of the same. "It's not turning back the clock; it's continuing on with something that's very capable of being there." More science, more good policies. There's nothing money can't buy, up to and especially when it's something meaningful to our history and our progress as a nation. While all of this may be true, and Gerrity's organization is poised to succeed, its success is built on the same principles that undid the bison in the first place.

Conni and Jason, for all their differences, share an alternate view of optimization not as a force of progress, not as a good process gone awry, but as a bad idea in the first place. Although Conni, in her appreciation of the cowboy, hearkens back to a later Edenic state than Jason's, hers is no less of a cry for restoration.

SOUTH OF THE AMERICAN PRAIRIE, NORTH OF THE TERRY RANCH, and a little to the west, I find myself back among the herd of thirty-odd buffalo that Jason stewards on Wind River reservation land. Ancestrally, bison, unlike cattle, were never fully domesticated by humans. Wild herds often migrated between various grazing grounds.

Jason is skeptical of the US government's stated commitment to bison. "We've been able to restore other species," he says. "Large animals like pronged antelope and bighorn sheep. But not bison."

I ask him why this is. His answer is simple and the same: "Ranchers and cattle. Antelope and sheep and deer don't compete with cattle. Individual ranchers benefit from cattle, whereas buffalo benefits the tribe."

He goes on to explain: a cow, or a buffalo farmed like a cow or propped up for show to attract tourists, is valued in dollar-denominated terms. Its value equates to what it produces or brings in. A herd of bison—stewarded by humans who not only divvy up its component parts but also value it unto itself, hold it up in stories and defend its environment—is a very different thing.

For Jason, getting to his vision is not merely a matter of time and hard work, though it involves that, too. All around, he sees obstacles. At Fort Peck, dozens of bison are in quarantine against brucellosis, a disease that affects cattle, elk, and bison. "They're sending buffalo to slaughter." Jason's voice is full of conviction, wavering with deep anger and sadness. To him it feels personal. Citing brucellosis is false reasoning, he says, as only elk can spread the disease to cattle. "Not a single bison has infected a cow," he says, whereas elk have. "Why are they slaughtering the buffalo and not the elk?"

The real reason, Jason says, is that buffalo compete with cattle. "We're living with imposed systems of land use and water use." He goes on to speak of the Shoshone's recent history: "If we didn't become ranchers, the land would've been taken away." Today, he tells, me, 86 percent of reservation land is consumed by range cows. Many of his neighbors don't even understand the importance of his mission. But to Jason, the stakes couldn't be more clear: "We're losing our buffalo, our elders, our language, our ceremony."

I ask Jason if there's anything that helps cut through his pessimism, that makes him hopeful his work will turn the tide. He responds with his unflagging conviction. "Over the years, I've had the opportunity to travel and speak with Indigenous leaders. Everywhere it's the same. A longing to get back to the truths we know." He goes on: "It may take fifteen or twenty, thirty years, but buffalo is how we heal. Heal everything." By this, he means the land, his people's food sources, their spirits, and their ceremonies. "We are Buffalo eaters. And we haven't been able to eat buffalo for 130 years."

Embedded in the treachery of optimals is a deception. An optimization, whether it's optimizing the value of an acre of land or the on-time arrival rate of an airline, often involves collapsing measurement into a single dimension, dollars or time or something else. In *Conservation and the Gospel of Efficiency*, Samuel Hays outlines the philosophical stances of early twentieth-century environmentalists, such as Gifford Pinchot, the first head of the Forest Service and adviser to several presidents on environmental matters. Pinchot saw productivity as the driving force behind conservation: "to make the forest produce the largest amount of whatever crop or service will be most useful, and keep on producing it for generation after generation of men and trees." This sentiment was echoed by the Massachusetts Commission on Economy and Efficiency, which wrote in 1912 that "the only proper basis for the protection of game birds, wild fowl and, indeed, all animals is an economic one."

One of Pinchot's contemporaries, John Muir, stood in philosophical opposition to this stance. Muir believed conservation was best done for conservation's sake; in the natural world, he saw beauty and even God. The two men butted heads, and they have come to be seen as representing opposing strains of thought in American environmentalism. Yet, for all their differences, the men also had something approximating a friendship. They camped together in Glacier National Park and worked together on more than one occasion.

Does Jason subscribe to Muir's notion of conservation for conservation's sake? When he says he wants more than productivity from the land, more than a token restoration, what does he mean? His answer is simple: he doesn't want to measure success merely in head of bison; he worries that when we assign a numerical value to each animal, we start to sink to the same language that wiped the bison out.

At the same time, Jason is a realist. When pushed, he acknowledges that things won't change all at once. The approach must be piecemeal. He knows it will progress slowly, through buying parcels, negotiating with neighbors, and weathering difficult seasons. An optimized system must be unwound bit by bit.

WHILE THE POPULATION OF BISON HAS CREPT UP CONSIDERABLY IN the past century, the same cannot be said of the ecosystem that supports them. Bison now roam on less than a hundredth of the land they once occupied, and this has an effect on other species and soil. They were once a keystone species; today, the majority are interbred with cattle. While the bison itself is not extinct, today it borders on what is called ecological extinction—that is, the disappearance of its range and the many interactions therein.

This ecological extinction is what Sean Gerrity seeks to remedy through the American Prairie Reserve and what Jason recognizes is at the heart of the problem. Routinely, conservation ecologists face these kinds of questions. Should a restored landscape target the original distribution of species, and if so, how is this achieved? Is it even possible to restore a tangled web of interactions that once existed? Restoration projects can fail because it's often harder to gain back diversity than to lose it in the first place and because we have already built infrastructure around the new state of affairs.

In some engineered domains, unwinding an optimization is as easy as flipping a switch. Tom Vanderbilt, in his book *Traffic,* describes a city that flipped traffic patterns in a single day, with everyone going from driving on one side of the road to the other. While a great deal of planning was involved, the switch itself was simple. The streets were repainted and signs flipped around. It took a few weeks for drivers to become accustomed to looking the other way at intersections. But there was only a single layer of rules and behavior to undo, and the switch was done in a synchronized and coordinated way.

Why is "flipping a switch" more difficult with many optimizations, especially those involving complex or nonengineered systems? Why is there almost always an asymmetry in putting the optimization in place versus walking it back?

One answer may lie in path dependence. Once an ecosystem has lived out a certain history, it can be hard to retrace the exact steps back to where it was. Unlike traffic rules, which are discrete and clear-cut, each step in an evolutionary path involves a certain amount of randomness, of unpredictability. It could just as easily have gone a different way. While randomness leads easily in one direction, to get back involves the precision of reverse-engineering those same steps.

Then again, not everyone agrees that going backward is always a good idea. Some critics have pointed out the problematic nature of restoration efforts writ large. Environmental journalist Fred Pearce, in his book *The New Wild*, is critical of a contemporary conservation movement that focuses too much, in his opinion, on restoring or preserving a specific species. Instead, he advocates for a more specific approach, wherein the entire network of species would be considered. Pearce admits this is difficult and without clear answers.

This ecosystems approach is not all too different from Jason's or Gerrity's. To Pearce, "alien" or invasive species, often seen as something to clear out of an environment in order to restore it, can also serve as part of the key to rewilding. In our now-optimized world, it's a myth to think we can simply unwind to some primordial past point.

What is often even harder to unwind than an optimized system is a way of seeing the world in terms of efficiency and zero sums, rather than as an evolving ecosystem with checks and balances. Jason has run into this in battling cattle ranchers on their economic terms, with their assessment of land prices. It's a language that's made its way into land use and environmentalism but stems from deep in our way of doing science and economics, increasingly predicated on specialization and measurement rather than open-ended search.

On the flip side, Jason's own language also espouses a romantic notion that's part and parcel of America's industrial history: to rebuild and restore what industrialization interrupted. In his book *The Machine in the Garden*, Leo Marx recounts a short story by Nathaniel Hawthorne, representative of a key feeling of this time. In the story, the narrator is interrupted from his reverie in nature by the sound of a clanking train. Hawthorne's tale paints mechanization as an intrusive force, piercing into an otherwise pastoral scene. After the interruption, the narrator finds it impossible to return to the Edenic state of bliss and oneness with the natural world.

Like the narrator's, Jason's pastoral state was interrupted in a major way. That his grief is mirrored by the sentiment of a nineteenth-century American Romantic author is perhaps cold comfort.

JASON HAS BEEN WORKING TO SECURE SOME SEVENTY THOUSAND additional acres, to grow the herd of buffalo on Wind River lands. The new acreage would be surrounded on all sides by two highways and two rivers, a protected island of sorts for the herd.

This hasn't been easy. First came smoothing relations with neighboring tribe the Northern Arapaho. Historically, they and the Eastern Shoshone are no strangers to conflict. Through some diplomacy, Jason has helped the Arapaho, who have eleven head of buffalo, create a counterpart to his own role for restoration. Now, the two are working together to grow the herd. Biologists say a minimum herd size of one thousand is needed to avoid inbreeding and loss of diversity; larger is ideal.

Then there is the issue of the lack of agriculture water covenants around tribal lands; the bison can't survive without water. And the private landholders adjacent to the land don't want to give it up or allow buffalo to graze. All this takes time.

Yet a glimpse into the future came in the spring of 2017, with the birth of the first bison calf on the reservation. In October 2020, another ten buffalo were released on the reservation, and in the spring of 2019, that herd grew yet again by five young bulls. And now, on the day of our conversation in the field, four calves were born in the early hours. Jason sees these as hopeful signs.

Some months after my visit to Wyoming, Jason and I chat via Zoom. He shares a slide deck depicting the shrinking boundaries of Shoshone land over the decades and the treaties that accompanied. There's no denying his deep sadness as he walks through the history.

As we speak, my camera freezes often, annoyingly. I apologize for my internet connection. "You must be out in the sticks," Jason remarks. I contemplate his and my relative stick-ish-ness and then say, "Yeah, I guess I am."

I do live out in the sticks, at least as far as geography is concerned. And by choice. It's hard to end up on a remote island by accident or chance. Yet I hesitated at Jason's comment. In other ways, I'm as connected as ever. Not just via flickering video calls but through an intellectual thread crisscrossing continents. I'd left behind the geography of optimization, but I'm still tied in to its imaginative landscape. Like the woman at the Terry Bison Ranch café, I hadn't counted on how hard it would be to get out of the place I knew.

Here again was the treachery of optimals. It's much easier for Magritte to go from pipe to painting than the other way around. It's far less simple to go from painting to pipe, from a policy of decimating bison to one of stewarding an ever-evolving ecosystem. It was less simple for me to return to someone else's past point of reference, to the slowness of the sticks, given the ground I'd tread in between.

This ground, I've started to realize, stretches out some way. I grew up in a Bay Area on the verge of the tech boom, in what would become a crucible for the mindset of optimization. I returned after grad school to work in the thick of it. And I come from a lineage of optimizers and engineers, many of whom made their way to America by dint of these skills. My mother earned a doctorate in operations research, and her one-in-a-million ticket out of Romania was abetted by optimization. As a top student in her field, she was sent abroad on a grant to complete her studies with the expectation that she'd return. Three years later, the grant was shut down, and getting out of the country got yet more impossible. My mother's allowed-for two years stretched into twenty, then thirty.

When in the late 1980s, she watched it all fall apart on our tiny television, for one brief moment, the possibility of returning was restored—at least in theory. In practice, returning was not a real option. For so long, my mother, and now my transplanted grandmother, as well, had lived with the impossibility of ever going back to the place they were from. There was a certain freedom, I came to understand, in this lack of choice.

A decade after the televised fall of Romania, when we did go back for a short visit, we found a place frozen in time. The presidential palace in Bucharest then claimed the second-largest building footprint in the world, after the US Pentagon. Through the 1970s and 1980s, the ruling Ceaușescus had frantically leveled the capital's historical buildings to expand their palace, as if to stop expanding would herald their decline. When on Christmas Day of 1989 they were cornered and executed by their own military, a fate flipped in one short week, time stopped. Now in the early 2000s, on my first visit to this place, I photographed cranes frozen in midair around the palace, holding cement blocks intended to keep building.

Of course, time didn't actually stop. History marched on, neighborhoods evolved, eventually the cranes were moved. Bucharest is a fluid and dynamic city, no more put on pause than a remote island with decent internet service can be considered the sticks. The frozen cranes and my frozen Zoom camera obscured these greater truths. And yet it's hard to deny the allure of each of these snapshots, each offering a tidy frame to aspire to or to leave behind.

René Magritte said of his painting: "My famous pipe! How people reproached me for it! And yet, could you stuff my pipe? No, it's just a representation, is it not? So if I had written on my picture, 'This is a pipe', I'd have been lying!"

And what a beautiful lie it would have been.

7

AFTER THE GOLD RUSH

AFTER OFFERING ME A SEAT, THE MAN ON THE TRAIN OFFERS UP HIS story of salvation. He's been playing dominoes with another passenger in the domed observation car. The one is weary of playing; the other, eager to talk.

I'm on the Amtrak Empire Builder line, which runs from the Midwest to the Pacific Ocean. At Chicago's Union Station, in the middle of an April snowstorm, I join a long line of families lugging suitcases, businesspeople with their folded papers, Mennonites and retirees and oil workers and military uniforms returning from a visit home.

The snow is heavy across Wisconsin and Minnesota. Most stops are a quick affair to on- and off-board; in Minot, North Dakota, there's a longer break. As we inch over the continent, a parade of characters unfolds. The domino man rambles on about religion. Two men go out for a smoke. A stout woman carries her tiny dog to a nearby patch of grass. The elfin sleeper-car attendant, on break, introduces himself as Michel Antoine Alexandre Timothé. I hear him introduce himself again, to others, always with all four names. He's of Pays Basque stock via New Orleans, where he's spent fifteen years researching a book on the New Orleans Basque population prior to the French Revolution—a story in which his relatives figure prominently and incongruously alongside French royalty.

At one end of the Empire Builder's route are some people we've met: the farmers and temporary laborers of the Red River Valley, Bob searching for his farm's legacy, Roger and Charlotte sifting beets at night and soaking their tired feet by their RV come morning. The sugar refined in the factories is loaded onto freight trains, and it, too, makes its way west (and south and east).

At the other end of the line is someone we have yet to meet: an agronomist who runs a bakery with his wife on a remote island in the Pacific Northwest—the island where I'm headed, where I now live. While most of the bakery's wheat and fresh fruit and herbs and eggs come from within a radius of five miles, one of their most important inputs, sugar, does not.

Here we chug across the continent, a continent of bison cleared, of ore and oil explored and exploited, of towns built up overnight. Glacier National Park glows as we wind slowly up toward sunset in the mountains. A linchpin of the Empire Builder route lies here in Montana, a tunnel cut through a pass, which allowed the railroad's first owner to build his empire.

Long before the Amtrak route got its name, Empire Builder was the name given to a man, that first owner and founder of the line: James Jerome Hill. We'll get to his story in a moment. It's worth pausing, first, to consider the founding myth of a blank-slate continent and its taming through ingenuity and engineering, with sweat and grit and brains and brawn. The myth that lies at the heart not just of American folklore but of optimization itself.

We saw in earlier chapters how this myth was built and how its promise has started to diverge from the reality it affected. We considered two responses. The first, to deoptimize optimally, is a response born of control. The second is driven by retreat: to unwind, to let go of the reins, to start anew. We also saw how each falls short. The first because it further entrenches an engineered optimization, continuing to cannibalize the very things deoptimizing looks to preserve; the second because it pretends away the optimized path, idealizing a past reference point or imaginary blank slate.

The mythos and mindset of optimization have been built up over decades, and our attempts to counter it have largely failed. This chapter asks: How do we reconcile myth with reality, and what comes next?

If Isaac Newton's attempts to reconcile alchemic mysteries inadvertently ushered in the era of optimization, James Jerome Hill might be seen as the person upon whom its curtain fell. Hill was born in Canada and grew up blind in one eye because of a childhood bow-and-arrow accident. His father died at an early age. Together, these tragedies shattered young Hill's dream of becoming a doctor but perhaps instilled an even wilder drive. By eighteen, he'd immigrated to the American Midwest, where he worked with railroads and steamboat freight. In the winter, Hill would bid on his own freight contracts, and by 1870, he and partners had started the Red River Transportation Company, running steamboats between Saint Paul and Winnipeg, right through what's now the beet-growing region of North Dakota.

In the economic panic of 1873, several railroad lines went bust. Hill, having built a small fiefdom already, now wrangled together business partners and enough capital to buy the Saint Paul and Pacific Railroad and negotiated rights from the Northern Pacific Railway. His ambition was to tie the West Coast to the Midwest, and the key was Marias Pass, just south of Glacier National Park.

There was a hitch. While the pass had been described by Lewis and Clark in the early 1800s, it had been unfindable since. Hill was determined to find it, as he believed it represented a superior route through the ragged mountains. Though known for scouting expansion routes himself on horseback, for this key mission, Hill hired a surveyor, whom he managed like a general dispatching his front line. The surveyor was able to find the pass described by Lewis and Clark, and Hill swooped in. This move allowed a hundred miles to be cut from the railroad's route. Its lower elevation meant tracks were less likely to freeze. About some of the nation's most gorgeous country, however, Hill was unsentimental: "We do not care enough for Rocky Mountains scenery to spend a large sum of money developing it." The scenery along the way didn't matter. What mattered was the most direct connection.

Now over the pass and through Glacier Park, the sun has set as the Amtrak train chugs on. Passengers pull hats over their eyes, lean their packs against windows to sleep on, cover themselves in outstretched sweaters or blankets they've brought along. A few linger in the café car with a coffee or drink, eyes and hearts searching for conversation, yet others retreat to sleeper compartments, where the soft chugging pulls them into sleep. There are announcements: Make sure you're in the correct car. Just after midnight the train splits in two, half the cars headed toward Portland, the other half to Seattle.

Here, on the West Coast, the train offloads its passengers and cargo and loads up again with a fresh set of faces. The dining car is restocked in the light rain. And then back it goes, rejoining its brother in Spokane, over the pass near Glacier, through the plains and the farmland and shimmying up against lakes and Milwaukee factories and gliding past tall office buildings into the Chicago late afternoon.

My own journey is far from over. North of Seattle, I make my way to my parked truck and onward to the island ferry. By now, I've lived here long enough to expect delays, and to appreciate the inefficiencies of ferry travel.

While the ferry system is just seventy years old, the islands have had their mainland tethers for some time. A hundred years back, a cannery connected Alaskan boats to the Seattle markets. Before that, these islands were the fruit basket of Washington State, barging out cherries, apples, pears, and plums to the mainland.

Today, most agriculture has consolidated to large irrigated farms in eastern Washington, the fisheries are in decline, and economic power has settled in tech-centric Seattle and San Francisco. The San Juan Islands are a cultural and political hinterland. And yet we're tied into the larger world in innumerable ways. The ferry brings in bagged bread and tourists from the mainland. The barges that once carried fruit now haul out discarded soda cans and foreign appliances.

The last step in getting home is also the least efficient. Locals like to joke, when leaving the island for a trip to the mainland, that they're "going to America." Indeed, the mainland is where much iconic modern-day America is found: Costco and Home Depot, IMAX movies and all-you-can-eat sushi, interstates and sprawling suburbs. These things are absent on our island, which boasts just two coffee shops and four restaurants, half of which dwindle to a few open days per week in the winter. There are two gas pumps outside the grocery store, and it's about a dollar more per gallon than in "America." Many try to stretch a tank until they make it off, scurrying to fill up between a medical appointment and a stop at Petco. As one might expect, and as is advertised to the tourists who like to bicycle here in the summer months, there's not a single stoplight on the island, either.

The ferry boat itself is the result of a highly optimized means of production and systems integration. But to be a passenger on the ferry necessitates patience and calm. A ten-mile journey takes forty minutes at best, and that's not accounting for on- and off-loading, nor for the frequent cancellations and delays. Because there are no reservations available to leave the island, it's a guessing game as to when one must line up, how many others might be there ahead, to make it on the boat. It's not unusual to wait two or three hours for a sailing.

Many complain about this system. It's full of headaches and slowness. And yet most people who are here have chosen it, in one way or another, not only the ferries but the full set of inconveniences and costs associated with living on an island.

I think of the ferry as the counterpoint to the tech cruise off Florida, and not just because both involve big ships. The Florida cruise created artificial slowness, in service of an optimized cause: maximizing onboard "networking." Everything about the cruise was sleek and perfected: the schedules, the food, the decks and swimming pools. The ferries here are an older fleet, inherently slow and prone to breakdowns. Their posted schedules, and more important the human schedules they mediate, have slack built in. They are also places of connection, where individual routines butt up against one another, sometimes resulting in new friendships.

ALTHOUGH RAIL PASSENGER SERVICE ENDS SOME THIRTY MILES BE-fore the ferry terminal, the railroad track continues almost to America's western edge. At this end of the line, on the other side of the continent that Hill pieced together, are some people who decided to hop off the continent. Nathan and Sage are two of them.

I find Nathan in one of his fields just east of the island's small lake. His and Sage's bakery, Barn Owl, uses grain grown on their own fields as well as by other farmers on the island. The present field also comprises Nathan's test plots. He's been working with the agricultural extension school in the Skagit Valley to test the performance of commercial, organic fertilizer versus homemade compost versus no fertilizer at all. Results indicate that wheat grown with homemade compost beats the other two, which perform about equally.

Nathan has also been experimenting with a set of ancestral wheat varietals. Over the millennia since we first began cultivating grain, humans have coevolved with plants: maize got bigger and wheat shorter, fruits were selected for greater sugar content and tougher skin, and even our gut bacteria evolved as we became farmers and husbanders of livestock. Then, with Norman Borlaug's green revolution, things started to get out of sync. One of Borlaug's inventions was breeding a shorter wheat, about a three-foot stalk, that was less likely to fall down in the wind than the six-foot varietals grown before. A short stalk could also carry more berries, resulting in more grain per acre. Harvesting machines had an easier time harvesting the grain closer to the ground, and new gas-powered harvesters were developed, some up to thirty feet wide and weighing some fifty thousand pounds and able to harvest nearly a hundred metric tons of wheat per hour. At current prices, that's several tens of thousands of dollars per hour, from a machine that sells for half a million dollars and costs thousands per hour to operate.

Nathan is going in the opposite direction. His wheat is six feet tall and harvested by hand or a harvester a fraction of fifty thousand pounds, threshed on a mechanical thresher. Then it's carted across the street to a neighbor's property, where it's milled on a device reminiscent of a large church organ, brought out from Pennsylvania years ago.

Although Nathan grew up off the grid in a cabin in the West Coast mountains, he hadn't given much thought to his food sources until his mother gifted him and Sage a copy of the hippie tome *Nourishing Traditions*. They read the book on a long drive down Interstate 5 to California.

The couple had been hankering for "a life path that had purpose," and food seemed to be it. They settled in Berkeley for a couple years. Sage got a job at a bakery, Nathan studied landscape architecture. After a while, they moved back to the Pacific Northwest. Nathan scoured the USDA seed banks for varietals he thought might be suited to the area, looking in particular for geographies with deep wheat-growing cultures, and partnered with local growers to cultivate them. All the seeds that succeeded ended up being from pre-Green Revolution varietals.

"It's funny," Nathan quips. "We work in this chaotic system; we try to harness it. And yet the more futile it seems, the more I realize there's a natural law, that every molecule is used efficiently."

Throughout these chapters, we've been skirting around an idea. The desire to connect, to harness, to control—the drive to optimize—is often met by its opposite: the desire to escape a set of constraints or strictures. James Jerome Hill was driven to connect a continent but couldn't do so in a vacuum. He needed places and people to tie together. He needed Wall Street's financing as much as he needs the Scandinavian immigrants to populate the railroad towns, the sugar at one end and the bakeries at the other. To build the bridges, he needed the islands.

Nathan is likewise tied in. On an island, spiritually and physically, he still relies on the railroads and ferries to get his goods. He still relies on farmers like Bob, and scientists in the vein of Norman Borlaug, to produce his sugar.

It is this codependence—of islands and railroads, of the twin impulses of control and retreat—that is at the heart of why optimization grabbed hold and took over. It's also at the heart of what comes next.

ISLANDS HAVE LONG PROVIDED THE BACKDROP FOR ALL SORTS OF case studies—for the stranded individual and shipwrecked family, for planned utopias, for the evolution of species.

It's hard to get to an island, but once a species successfully arrives, it can be easier to take over, compared to a large and filled-in continent. So goes the idea of island biogeography, a branch of ecology first developed in the 1960s with the goal of explaining the varying composition of species on different islands. What purpose do islands serve? Why are some islands so diverse, while others are relatively sparse?

The theory posits two main factors to explain how an ecosystem comes to occupy a space: the rate at which new species colonize the island and the rate at which existing species go extinct. If a new volcanic island arises near a mainland with hundreds of types of birds, some birds will find their way. A full island is generally harder for newcomers to colonize than an empty one. A long distance from the mainland implies a lower rate of immigration. Large islands tend to have more possible environmental niches, more species, and lower extinction rates.

The Indonesian islands of Krakatau were decimated by a volcano in 1883. Since then, they've proven a fruitful place to gather data. At first the islands saw rapid recolonization of species from nearby islands. Then the rate of growth in diversity began to slow. In more recent years, they've provided lifelines for species being pushed out of more competitive mainland environments.

Islands such as those of Krakatau let us observe a nonengineered optimization process at play. They can also showcase the ways in which engineered optimizations, such as increasing connectivity from railroads or internet, play out differently. In these ways, islands are reminders of the tension between interconnection and insularity. In order to reconcile the fruits of our era of optimization with its shortcomings, in order to see what comes next, we can't ignore this tension.

The poet John Donne famously reminds us that no man is an island. These days, with railroads and ferries and airplanes and the invisible span of the internet, it seems that no island is an island, either. Islands are becoming ever more connected to the mainland, or as Donne put it, they're a "piece of the continent, part of the main." A shift is thus needed in how we think about efficiency, about the importance of local optima. This means neither a return to some preoptimization golden age nor the mantra of progress at all costs.

ISLANDS CAN ALSO SHOWCASE WHAT WE'VE LOST. IN LATE 1918, THE steamship *Talune* arrived in Western Samoa from Auckland. Several of the sailors aboard were ill, so as they went ashore, the captain ordered them to hide their symptoms. The island was newly under New Zealand jurisdiction and an important asset for global trade. The captain had a schedule to make and goods to unload. In the months that followed, Western Samoa was decimated by what came to be known as the Spanish flu. The island lost nearly a quarter of its population, many of them healthy adults in their prime. Meanwhile, on the neighboring island of American Samoa, not a single life was lost to the pandemic. What did these two colonies do differently?

American Samoa's officials had gotten wind of the pandemic that was sweeping the world. They also had just a third of Western Samoa's trade. So when American Samoa ordered all ships to quarantine for

five days off its coast before anyone set foot on land, captains com-
plied, or they mapped other routes. The island hadn't made itself a
vital stop for trade.

American Samoa's quarantine meant that some commerce went
elsewhere, and some ships sat offshore. By the time the virus made
landfall, it had become far less lethal.

On the other hand, Western Samoa was a key stop for optimized
global trade, turning its isolation from a strength, in terms of its
distance from disease, into a weakness as it became a petri dish for
infection. Trade routes in the South Pacific were built over many
years and countless negotiations; by the time pandemic hit, it was too
late to unwind them thoughtfully or at all. Instead, Western Samoa's
population was wiped out in mere months.

Optimization works under constraints. The geography of an island
offers such constraints and, in the case of Western and American
Samoa, a demonstration of how globally connected we are. Islands
allow pockets or niches of species ecosystems to evolve without giant
predators, insulated from disease and parasites. An island left insular
can thus help revive the mainland in times of trouble.

Conversely, mainlands are important to islands: they provide the
pioneer species that populate islands in the first place, and they allow
ecosystems to evolve over larger swaths, thus generating the diversity
that eventually lends richness to islands, too.

One of the consequences of optimization has been a breaking
down of the borders that typically make islands, islands. With optimization, we've preferred the mainland, and the mainland way of
seeing, as a vast blank plain to be optimized on, rather than an un-
dulating landscape with pockets of rejuvenation.

IN 1910, NOW IN HIS SEVENTIES, JAMES JEROME HILL PUBLISHED AN autobiography, titled *Highways of Progress*. It begins with a simple statement: "Nations are like travelers." Hill goes on to lay out plans for restoring what he sees as declining productivity in America. We'd reached our borders, expanded sea to shining sea, decimated our most important natural resources, and plucked the low-hanging fruit. Hill predicted the total depletion of coal by 1950, bemoaned a decline of soil quality, and proposed steps to double the yield of grains in Midwestern fields. Things wouldn't come so easily henceforth. Industriousness and invention were required.

James Jerome Hill lived through and embodied a moment of expansiveness. A 1910 photo shows him, bearded and dapper, walking arm in arm with two of the decade's most powerful bankers: Charles Steele of J. P. Morgan and George F. Baker Sr. of the First National Bank of New York. When Hill retired from his position as chairman of the Great Northern Railway, his farewell letter made clear his position in history as well as the spirit of the times: "Most men who have really lived had had, in some shape, their great adventure. This railway is mine."

The historian Frederick Jackson Turner has written that "to the frontier the American intellect owes its striking characteristics." When the frontier was closed, he argues, an era of history ended. As Hill reflects in his autobiography, capturing resources on an untapped continent looks a lot like increasing productivity within a closed system. The two, however, shouldn't be confused. When Hill looked out as a young man, he saw a world unbounded. By the end of his life, his transportation empire had drawn the boundaries, and Hill left to the next generations the mandate to squeeze still more from that which lay inside.

Hill's legacy, like his life, looms large. His biographer Albro Martin writes that not long after Hill's resignation from Great Northern, with the US entry into World War I, twentieth-century America "with all its international obligations was born." Martin adds that "James J. Hill was one of the midwives" of this new era. In a novel that captures the budding age in all its grandiose hope and splendor, F. Scott Fitzgerald's Henry Gatz says wistfully of Jay Gatsby: "He'd of been a great man. A man like James J. Hill."

Alongside Hill, the railroad itself has long served a gorgeous mythos, as a backdrop to the cowboy stories of John Wayne, as a passageway for Florida oranges into New York winters, as the conveyance of Scandinavian settlers to develop rich Midwestern soil. Just as James Jerome Hill's life coincided with the railroad's expansion, with the heyday of optimization, so did his death herald its decline. In the years following, the United States broadened its reach in the world, increasing its reliance on grain from abroad. Global commitments grew in complexity and scope. The fifty years after Hill's death marked some of the most stable and remarkable growth in America's economic history, fueled by these entanglements as well as productivity at home.

Everything was great, or so the story goes, until it wasn't. Economist Tyler Cowen has called the decades since the 1980s "The Great Stagnation." We've picked off the low-hanging-fruit of fast growth and now face a stasis. Not unlike Hill a hundred years ago, Cowen frets about the nation's decline in productivity. Except now that the fruit is gone, only the myth remains.

THE MYTH OF OPTIMIZATION PERSISTED PAST ITS PARTING WITH RE-ality, and for me it persisted long after I boarded the ferry off the continent onto this island. Unlike Nathan and Sage, I haven't fully hopped off, nor opted out. In the years since buying the run-down cabin, I fixed it up, I built connections, I learned skills. I took the train across the States, then drove east again two years later, and each time found my way back home.

I jokingly ask Nathan when Barn Owl will start selling its bread at Walmart. He laughs and then responds earnestly: "Never." Indeed, everything about Nathan and Sage's bakery is earnest.

I'd likewise moved away from optimization in earnest, by moving here. And yet, I knew my move was contingent. In the same sense that Nathan and Sage depend on the railroads to make an island bakery possible, I hold a debt to optimization as a way of seeing, for giving me something to believe in and ultimately opt out of.

I'm not alone in this. Many are grasping for something new, without fully knowing what. For many of us, we know only the metaphor and its opposing argument. You can opt in or opt out. In between is only limbo. An article in *The Wall Street Journal* describes the re-cent phenomenon of "quiet quitting," or doing as little as possible to maintain one's job and nothing more. A young man elucidates the concept: "You're quitting the idea of going above and beyond . . . no longer subscribing to the hustle-culture mentality that work has to be your life." The article explains that many young people "reject the idea that productivity trumps all; they don't see the payoff." Even opting out is anchored by the framework of efficiency. One can ask only, What is the payoff?

Of course, optimization as a scientific tool kit hasn't died, and it will continue to prompt advances in the years to come: in medicine and biology, in scheduling and logistics. But as a way of organizing the world, its dominance is waning. For too long, we've tried to ignore our debt to the world James Jerome Hill wove together, and that debt is now coming due.

The metaphor of optimization is showing ruptures precisely because of the paradox at its core: the more we optimize, the less we're able to do (and see) things any other way. Put differently, we can't reconcile the shortcomings of a particular optimization within its own bounds . . . and likewise: we can't reconcile the shortcomings of the idea of optimizing within the framework of optimizing. Like the railroads, it's a centripetal and unifying force.

Is there any way to resolve this paradox? For better or worse, the heritage most of us reading this book "belong to" is that of progress, rationality, and optimization. It's bestowed many gifts, along with its costs. And while this way of seeing is breaking down, it's unlikely we'll do away with it wholesale. The preceding two chapters presented two possible, but flawed, solutions. We can double down and optimize more, or we can seek to unwind, in search of some past reference point. Both, in different ways, reinforce the broken lens.

In leaving San Francisco, I was looking for a way out. I wanted to better feel a connection between my own actions and their effects. I wanted to frame a wall and feel the wind subside. What I was yearning for, in other words, was a reclamation of those things lost to optimization. This yearning is not uncommon today and is increasingly rearing its head in all sorts of ways.

One way we see this yearning expressed is in a search for a less frenzied way of life. In the increased preferencing of aesthetic over purely material or economic ends. Take, for example, the "experience economy," the millennial trend of going after intangibles in lieu of things—skydiving over a Rolex. Although both are driven by consumerist acquisitiveness, and the trend is often dismissed as empire-in-decline-type status signaling, it's also a sign of something deeper. The former, the experience, lends itself less easily to pure quantification, pure comparison, pure optimization. Its value lies in the subjective understanding of the participant, rather than the judgment of the open market.

This yearning is seen not only in "quiet quitting" but in a genuine slowing down, a step off the hamster wheel of piling up wealth and climbing ladders. It's seen in the increasing despair of those who cannot make ends meet, who know that something has to give. It's apparent in the farmers who are turning away from the servitude of global markets, in conservation efforts that move from sterile bean counting to genuine stewardship.

The yearning can also be seen in a turn toward the tangible, the specific, toward a reclamation of place. It can be felt strongly in recent years as populist movements gain traction against philosophies of centralized planning and government, as restaurants emphasize local food, as more young people grow tired of living in anonymity and seek a tangible community, whether rural or urban, secular or religious, or ethnic or political. It's present in Nathan's focus on growing wheat on his postage-stamp farm. In Aisha Nyandoro's drive to help mothers in Jackson. It's even evident in corporations as they acknowledge the fragilities inherent in our current system and work to regain robustness and, too, in national politics that are turning increasingly away from global entanglements.

This decoupling does not mean detachment. A recent *New York Post* article asserts that off-grid living is no longer just for hippies and fundamentalists. "It makes it sound as if we are just withdrawing from society as we know it. No, it's the opposite," a man named Norm Vaux proclaims. "A lot of people like me are being pushed toward something that better aligns with their values. I'm as connected to the world as I ever was."

The yearning is expressed, too, in a search for greater integrity between part and whole. A basis in first principles, a recalibration with and attention to scale. Sociologist David Graeber describes, in *The Utopia of Rules*, the safety we find in systems and bureaucracies. Increasingly, we seem willing—and, in some cases, forced—to give up some of that safety in order to gain coherence. An understanding of where our money goes and where our products come from. Amid the mire of abstractions and video games and antidepressants, there is a craving for something real.

These various expressions of yearning are a far cry from those that characterized the peak age of optimization. Optimization grew by commoditizing, flattening, relegating choice from the individual to the system—all while providing the illusion of "choice" in the form of endless supermarket varieties, of all-you-can-eat news articles, of identical packets of commodity sugar, sifted into a hundred differently branded bags.

Jane Jacobs writes: "When our own society was much poorer than it is today, it nevertheless managed to meet the inherent expenses and inefficiencies associated with its continuation. How did it do this? How do poor but vigorous cultures manage their continuation today?" Ironically, optimization has given us not more choice but false redundancies. Jacobs answers her own question like this: When we let go of control, we're able to ·egain some slack and place, or the "varied individuals who have varied ways of fitting into the culture and contributing to it." The redundancies and insularity of small pockets, in other words, is what contributes to the vigor of the whole.

These days, islands and mainland are interconnected in myriad ways beyond slow ferries. People and goods and diseases do make their way over, as they always have, and information is conveyed back and forth at record speeds. Still, local customs and inside jokes persist here, inured to the mainland world.

To reconcile optimization with its shortcomings and losses requires forgetting neither islands nor mainland. When information is incubated in small pockets, when products are cultivated and crafted in tight circles, marvelous things occur. Yet leave those circles alone too long, and they become stagnant, echo chambers, or worse. Likewise, the mainland can produce enormous quantities at impressive speeds, build skyscrapers, and feed armies, but too many years of noise and speed and clamor will leave even the most hardened city dweller longing for a little quiet, for a corner to himself, an escape route out of the machine.

— — —

ALTHOUGH THE RAILROAD TRACK REMAINS, AND TRAINS STILL travel this stretch of continent, there came an end to the Empire Builder era. By the turn of the twentieth century, two factions were at odds for the northernmost railroad lines. Hill, aided by J. P. Morgan, controlled the Great Northern Railway as well as the Northern Pacific. Another faction owned the Union Pacific Railroad: Edward Harriman, who was backed by William Rockefeller.

The core territory contested was control of access to Chicago and the trade that came out of it. Harriman hatched a plan to quietly buy shares in Northern Pacific, with the aim of wresting control of these lines. He was on the verge of enacting this plan, when Hill found out and alerted Morgan, who declared war. Morgan and his associates bought all the shares they could, cornering the market and driving the price of Northern Pacific from under $100 to more than $1,000 per share. Hill and Morgan battled Harriman and Rockefeller, with neither side able to gain enough shares to fully control the lines in question.

This battle fit the mold of magnates fighting over territory, but for one hitch. For several years, the market had attracted speculators, many of whom had sold Northern Pacific short. Not only would the speculators be ruined if the stock price remained high, but the entire market faced collapse.

Eventually, the two sides settled into an unhappy agreement, which gave Hill and Morgan control of the contested lines. Their victory, however, was short-lived. In 1902, newly elected president

Teddy Roosevelt dispatched his Justice Department to break up Hill's conglomerate, and in 1904, the Supreme Court ruled to dissolve it. The immediate effect was a drop in foreign trade. Hill was no longer able to maintain competitive rates in distribution, and trade with Japan and China declined by 40 percent. Everyone, or so it seemed, had lost.

James Jerome Hill's funeral was held on May 31, 1916. At precisely 2:00 p.m., every train on the Hill lines, wherever it happened to be, paused on the tracks for five minutes. Hill's biographer concludes, of not just the man but his era, that "abundance has always been the exception, and not the rule, in the human story." This exceptional abundance was, in part, what created men like Hill—men who "when we need them again . . . will be found walking in our midst."

Just as the railroads interrupted Hawthorne's peaceful reverie, just as they served as the signal of a new era in Western movies, a new set of policies had interrupted the railroads and their promise of eternal abundance. Or perhaps it was simply that they'd reached the West Coast, and there was nowhere left to go.

IT'S BEEN SEVEN YEARS SINCE I VISITED BOB IN NORTH DAKOTA, TWO years since he planted his first crop of GM beets. In those same years, the concern over GMOs at the supermarket has faded. Fewer people seem to look for non-GMO sugar anymore or to distinguish between beet and cane. I ask a commodity-trader friend, who's spent a decade on sugar contracts, what happened. For a while, the trading desks looked with a keen eye to the price differential between organic and conventional, with GM crops falling into the latter category. The two are still treated as separate, she says, but the idea of a watershed moment is gone. What changed, I wonder. Her answer is nonchalant: "People stopped caring." A 2021 headline in *The New York Times*—"Learning to Love G.M.O.s"— confirms this sentiment and sings the praises of produce engineered for nutrition and consumer tastes.

Maybe that was the problem all along. When the seed companies optimized for ease of growing, for farmer-facing traits, consumers weren't impressed. Now that they're optimizing for things consumers care about, the pushback has subsided. Never mind the health concerns, the loss of crop diversity. It was only a matter of changing the objective function.

Now I ask Bob how Roger's doing. "Well, Roger just disappeared," says Bob. "I mean, I know where he is. He's in his trailer on his patch of land in the backwoods of Missouri. Charlotte, too. But he disappeared. Stopped coming back."

I, too, know where Roger is, because I'd tracked him down once before, on that same patch of dirt in rural Missouri. It was 2015, the same year I attended the beet co-op shop meetings in the Red River Valley, the same year I spent long afternoons in conversation with Bob, the same year my romanticism about mathematical modeling began to wane. Missouri is a long humid drive and a far cry from Grand Forks, and Roger's muggy driveway winds on forever. I'd made it, along with a friend and a second camera. We thought Roger might shed some light on the life of temporary workers at the beet factories; we thought his wisecracks would lighten the mood. It was high summer, still and hot with only the buzz of flies to break the muted air, a lull before the rush of harvest.

We had a nice chat, the four of us, just long enough. Roger offered us iced tea and made us laugh. He didn't have much to say about the factories or the RV workers. Those things were, when it came down to it, just life. The afternoon stretched on and then we drove off into our separate worlds.

8

BABYLON

THIS STORY BEGAN WITH A BULLDOZER CARVING RUTS INTO A FIELD.
It dove down to the plains and up to the high desert. It traversed a
continent and boarded a ferry, and now it comes to rest on an island.

Isaac Newton wanted to harness the mystical and material forces
of alchemy. He labored for decades toward nature's secret fire, in
search of the green dragon and the philosopher's stone. Then, as he
became disenchanted or bored or simply grew old, his interest waned,
or perhaps it went underground. The century turned toward a separation of experimental science from arcana, and Newton to his cushy
university post. Ultimately, Newton never found the philosopher's
stone, and while some alchemical research made its way into his later
work on optics, it's not what he's remembered for today.

Newton, "the last of the magicians," looked backward to alchemy
and inadvertently ushered in the age of optimization. If Newton is
the first of this era's pioneers, James Jerome Hill, empire builder and

"midwife to the 20th century," might be considered its last. Hill looked forward, not backward, and yet his life's work consolidating a railroad empire helped close a chapter in America's history. The mythos and practices of optimization, of course, flourished long beyond Hill, through Stan Ulam's computer simulations and Allen Gilmer's geologic mappings and Marie Kondo's efforts to tidy up. But Hill was the one to plant the proverbial railway stake. He connected a continent and consolidated a system of commerce and, in so doing, paved the way for optimization's rise and ultimate decline. As the boundaries were closed, the fences fortified, optimization as a technology excelled. As an epistemology, however, optimization was starved of outside influence and could only cannibalize from within.

That same impulse, to close in on a finished frame, to capture and understand the space inside, drove me across the continent and back. A mathematical model is not too different from a story. Whenever we frame something, all that lies outside our tidy rectangle begins to blur.

Indeed, it was hard for Hill to see past the era he helped bring to fruition. The question of what's next is rather left, unresolved, to Hill's beneficiaries and debtors. How can we reconcile the fruits of optimization with its losses? How can we rediscover slack, place, and scale? How can we differentiate the islands that efficiency has sought to merge? And how can we honor the mainland efficiencies that populated such islands in the first place?

If the previous chapter considered this question with an eye to the material world, to agriculture, economics, and trade, this chapter asks the question for the individual. How can we reinhabit the perspectives that optimization cannibalized? It's improbable, not to mention impractical, that we all move to islands and begin to farm local fields, as Nathan and Sage have done. But the coming age requires that we rediscover islands of knowing and, with them, a more human scale of understanding.

In North Dakota, we ran into a certain sadness, a malaise, in Bob's quest to disengage from the wheels of progress, a sadness that's shared broadly across America today. The feeling that unchecked growth isn't serving us and that in the face of optimization's many gifts, its cheap produce and lofty towers, we've lost a certain agency. We've also seen throughout these chapters a set of twin urges: to alleviate this malaise either by doubling down on efficiency, as Sam Altman and his technologist ilk propose, or by attempting to escape or unwind it altogether, as Jason Baldes hopes to do.

The problem with each of these avenues is that each perpetuates the primacy of optimization: the first by entrenching it as a way even to deoptimize and the second by shoving all our chips from the present onto a past point of reference.

If neither doubling down nor escape is the way, what, then, is next?

IT'D BE NICE TO BELIEVE THAT THE FUTURE LIES IN NORTHERN KENtucky. When Amazon's Air Hub project broke ground in 2019, Jeff Bezos himself hopped into a front loader and dug at a ceremonial pile of dirt. "Let's move some earth!" Bezos, in aviators and a tucked-in white shirt, cried to a small cheering crowd of executives before cantering out to the camera-ready John Deere.

When I visit in 2020, I watch the man on the bulldozer and wonder at his thoughts. Does he think about the layers he's cut through, the dirt displaced? Did someone in his family, his grandfather perhaps, work the land? Did a side job as a mechanic pay the bills? Maybe his mother and aunts eventually sold off what remained of the farm. It could well be I've got the story wrong. Maybe his father was a night manager at the coal plant. Does he now think, every time he drives by the river, of all that water, cooling the factories where his father worked, quenching the fields? Maybe he imagines me as I imagine him, lumps me in with all the fancy coastal people, blithely filling our white countertops with espresso machines assembled in Italy from Chinese parts. Maybe he thinks now about his own bulldozer, about the parts that need replacing, or the articles in the newspaper describing the jobs the Air Hub will bring in, about the history it will push out.

Amazon's Air Hub, after all, is being built on ruins. Beneath the bulldozer operator's own are generations more, of settlers, of the region's first inhabitants, of the land before it was peopled. And the ruins go back further yet, to the stories we still tell ourselves. In Babylon, one story goes, the lottery begins as a simple game of chance. Tickets are sold, winners are drawn, prizes awarded.

As Argentine author Jorge Luis Borges describes it in his short story "The Lottery in Babylon," the game soon evolves. Punishments are doled out alongside prizes. Eventually the Lottery becomes compulsory. Its cadence increases, until the outcomes of its drawings come to underpin every element of daily life. Mundane events and life turns are subject to the Lottery's hold.

In the Lottery's early days, you might win a bottle of wine or a new home. As time wears on, the stakes go up. Fates are reversed. The bottle of wine is taken from you, the homestead. One day you are made a rich tradesman, the next a beggar. You could receive a death sentence or be crowned a king.

The Lottery is run by a shadowy group of operators called the Company. While at first it serves as "an intensification of chance, a periodic infusion of chaos in to the cosmos," as the Lottery's reach expands, it begins to control not just outcomes but how one's fate is understood. Over time, belief in the Company's explanatory omnipotence eventually cannibalizes all others.

The narrator describes it all with unabashed awe: "I have known that thing the Greeks knew not—uncertainty . . . Like all men in Babylon, I have been pro-counsel; like all, a slave."

Still, cracks begin to show. Historians must rewrite their tomes or leave them open to a branching of possible outcomes, around not just the future but the past. Knowing full well these cracks, the narrator doubles down on his faith. Whether it's an earnest faith, or one tinged with irony, is left ambiguous. We never know how he feels. Either way, his is the language of the Lottery's centralizing force.

Borges, born in Buenos Aires and overtaken slowly by a blindness that became complete in his midfifties, reminds us of the oppression of a singular way of seeing, an engineered control over the world that both alleviates and accelerates anxiety.

It's not unlike our present-day malaise. It's what we feel when faced with the promise of the best and the fear of choosing any path that may be less than that. We're wont to claim, nervously, that some outside force like the Company controls it all, anyway. It's the urge to make the most of the system, to get rich quick, to win big—and at the same time to crumple up calendars, gather up our children and loved ones and a few possessions, and move to the deepest woods. It's the malaise that coaxes us away from any drastic measure, to pencil in instead a couple "mental health" days off throughout the year, to remind ourselves to meditate and take the stairs and keep things in perspective. After all, there's only so much we can control.

There's only so much we can control. This is the uneasy middle ground between two impossible extremes: the drive toward total conquest and the drive to retreat. Our malaise takes root in this static middle.

No one of us can be the best for very long, and for most, opting out is difficult. We can't truly leave behind the world of optimization, or maybe we don't want to. Our livelihoods, our quality of life, our social connections, and our very way of understanding the world depend on it. It's a feeling that, as at the Hotel California, "you can check out any time you like, but you can never leave."

As he tells his tale, we encounter the narrator of Borges's story on a ship bound out of town. Whether his departure is a testament to the Company's hold or to its limitations remains ambiguous. "The Lottery in Babylon" recalls this paralysis, amid an ever-growing culture of efficiency. We feel its breakdown, but also a powerlessness to define what's next, to drop the current script.

THROUGHOUT THESE CHAPTERS, WE'VE SEEN HOW MATHEMATICAL ideas take cultural shape and vice versa. Claude Shannon's quantification of information allowed us to atomize and harness uncertainty, John Stuart Mill's perfect society became expressed in the equations of economists, Stan Ulam's game of solitaire framed the lens of probability, and Marie Kondo's early inspirations created the mechanics for organizing our material reality.

There are those who with mathematics seek to keep alive James Jerome Hill's empire building, to come up with a still more unified theory to reconcile optimization with its shortcomings. This is what Zappos sought, what Sam Altman is preaching, what Herb Simon experimented with, what even antinuclear activist Ed Grothus evangelizes, if simply by rebelling against it.

And we've seen the opposite, as well, as illustrated in Borges's narrator: a nagging doubt, a desire to escape. We saw it in Candide, who, faced with the Panglossian impossibility that we live in the best of all possible worlds, elected to tend his garden. We see it in Nathan and Sage, tending their own island fields. And we see glimmers of it in Bob and Jason, with their grand visions proudly poised against the grain.

It's what motivates back-to-the-land hippies and libertarian preppers alike. A traditionally American answer to the problems of a mechanized age. Off the grid and down a long dirt road. "If I could just get off of this L.A. freeway," mourns the gravelly voice of Texas songwriter Guy Clark. Wilderness offers the escape from industrialization. And it works the other way around, too: the archetypal Western movie has the clamor of the railroads providing a stark, inconclusive end to the pastoral blank slate of the frontier.

Yet this kind of ending, this escape, is less than satisfactory. As Leo Marx suggests in *The Machine in the Garden*, in American literature a retreat to nature is seen as the only tenable resolution to a world overtaken by machinery: "In the end the American hero is either dead or totally alienated from society." Indeed, he writes, "the resolutions of our pastoral fables are unsatisfactory because the old symbol of reconciliation is obsolete." In other words, the American fable lacks closure precisely because the wheels of progress interrupt and defer, rather than resolve, this malaise. Just as the bulldozer operator at Amazon's Air Hub pushes on into the field, interrupting and deferring and building what will someday be new ruins. Hewing to plans architected and stamped, insured and reinsured by the biggest in the business, nudged and approved at the highest levels of government, managed by the most trusted names.

FIVE YEARS PRIOR TO MY AIR HUB VISIT, IN THE SUMMER OF 2015, I returned from North Dakota to San Francisco. I'd spent much of my adolescence in the city, as an older layer of wealth was making way for a new tech boom. I came through again in young adulthood, as the city was coming into its own as a center of optimization. And here I was once more, on my way to a party.

As I approached on foot, it struck me that at the top of this same hill, I'd been invited more than a decade earlier, by a friend of a friend, to a very different gathering. It was the late 1990s, the hosts the teenage children of the old guard. The old guard, before tech permeated the city, meant actors and mystery novelists and a handful of minor scions of West Coast wealth. My friend showed me into the white stucco mansion, walls adorned with modern art, carpets plush, chandeliers glimmering. There was a button to call security, another to call the maid. The garden outside was trimmed in hedges and ivy and lush with evening fog. I hadn't yet traveled to the East Coast, or read *The Great Gatsby*, so if you'd asked me in 1999 to define American wealth, this would have been my answer. I took it all in, knowing full well I didn't belong. There were fifty of us, maybe more, young and drinking in the late millennium, our glasses clinking loudly on the kitchen marble, unhindered, unsupervised. Perhaps the parents, the adults, clinked their own debauched champagne flutes in another wing somewhere, but no one knew or cared.

Somewhere in the intervening decade, the mechanical cranes bored into San Francisco, the high-rise condos shot up, the skyline itself transformed. A new kind of wealth poured in. And here I was, some fifteen years later, on the same block of the same hill, attending the after-party for a billion-dollar tech acquisition, at the home of the young CEO. The walls of his magnificent living room were mostly bare. The hallways were lit with sterile recessed LED. The party chatter hopped from one startup's valuation to the next, like a commuter train plugging through bedroom communities in late afternoon. Bored and curious, I opened the fridge, expecting all manner of delicacy. I found nothing but a case of sparkling water and the remnants of upscale takeout. Expensive whiskey was shown off in the cabinet. Two fancy cars in the heated garage. A generation of new wealth, living like squatters in the mansions of yore.

The whole scene disgusted me, in part because I understood I was complicit in its creation. I'd ridden the wave, partaken of the feast; I had little to show for it beyond some leftovers. All of us, in San Francisco and beyond, had failed to honor the previous era's mansions with refurbished decor. We'd also failed to take apart what needed undoing, to build something of our own. We were paralyzed.

I left around 3:00 a.m. and walked all the way back to my apartment, some four or five miles. The decision felt purposeful, although in retrospect I didn't have much choice. My dumb phone had no apps, the cabs were few, and most of the bus lines stopped running around midnight.

I left, but the story sticks with me. Just now, I listen to a similar feeling of despair from one of the fiercest promoters of optimization. In a leaked 2022 video, Starbucks CEO Howard Schultz reveals he's closing a large percentage of the chain's cafés. The violence has become too great. Baristas are feeling threatened. People walk in; they don't buy a thing; they use the bathrooms to get stoned. Meanwhile, Starbucks's drive-through and preorder business is thriving. People love the product, Schultz says in the video, yet the need for a third place, outside of work and home, that Starbucks once provided is now gone. What's left is work, and the homes where work increasingly takes place, and the optimized route from home to work to home.

Starbucks succeeded because it created a high-quality and predictable product and stamped it out across the nation. In 2021, there were about fifteen thousand stores in the United States, with locations as far-flung as Fairbanks, Alaska, and Guantanamo Bay. In many communities, the Starbucks café became the public gathering place. It was safe and affordable, the food and drinks reliable. Starbucks famously and effectively optimized its supply chains, and many smaller coffee outfits followed suit, dramatically increasingly the quality and availability of coffee around the world.

Schultz's leaked comments point to a growing bifurcation between the streamlined apps and supply chains on one side, and the human-scale need for community and connection on the other. The convergence point, these stamped-out third places, is what's broken. As we saw in the previous chapter, there's a tension and a codependence between these two poles, of streamlined and local, of control and escape. The one needs the other. Islands need the railroads, just as railroads need the islands. This has allowed a certain coming-to-terms between global supply chains and local bakeries. But how does this translate for us as individuals? How might we, paralyzed by optimization and faced with its breakdown, resolve this tension and find a new middle ground?

My hunch is that a way forward involves admitting this paradox: that we can't have escape without control, nor control without the possibility of retreat. We can't move on until we acknowledge the mansions we're squatting in, the layers of ruins upon which our Air Hubs are built. Our job now may not be a full reconciliation per se, nor a demolition, but rather one of choosing how to see.

ISAAC NEWTON GOT OLD AND FAT. HIS HAIR WAS DISHEVELED, HE engaged in gratuitous quarrels with colleagues and strangers alike. He eventually looked away from alchemy.

Historian Thomas Kuhn tells us that big transitions all begin with small irregularities that emerge in the present theory, first quietly, then with more fervor. To Kuhn, this first bloom of anomalies is the most important—not the upheaval and revolution that follows, not the later consolidation of a new theory, but that moment when the pieces don't quite seem to fit.

Of course, ushering in a new era was never Newton's plan. Some biographers see in his work the drive for unity, for a synthesis of domains. Others, such as William Newman, consider their subject rather "an out of control genius," a man with deep curiosity and a sort of Midas touch for scientific insight. In this view, Newton wasn't seeking unity or reconciliation, but stumbled into it by ferreting after loose ends of alchemical writings.

Part of Newton's fear, as well, was that admitting the gaps in synthesis, the anomalies, the places where alchemy's arcane formulas could not be fully reconciled with experimental science, would invite in a sort of atheistic vacuum. While Newton himself had deep—and for the time, heretical—questions around his faith, he disparaged the kinds of reductionist thinking, as, for example, Descartes proposed, that would remove divine influence as one leg of the explanatory stool. The stool may have been lopsided, it may have rocked when you sat down, but kicking out a leg would cause the whole thing to collapse.

The century since James Jerome Hill has left us with the kind of vacuum Newton, for all his personal disbelief, so feared. At the same time, as our optimizations have tightened, our reliance on a machined and deterministic world has only grown, as has the mythos of efficiency. We reduced things to the denominations that optimization could measure and, in so doing, wobble between the extremes of believing whole hog and believing in nothing at all.

A similar dilemma paralyzes the narrator of Borges's story. He can't control the Lottery, nor can he escape it. He can only, on his ship bound out of town, speak to the Company's all-consuming power, in the language the Company created.

But then again, that's not the end of the tale. His isn't the last word.

Centuries later, we know the history by heart. Babylon as a place fell apart. Babylon as a story endures.

PERHAPS, THEN, THE REMEDY IS NEITHER IN DOUBLING DOWN NOR in escape, but in choosing how to tell the tale. This choosing requires a certain faith. It requires opting out, not in the sense of escaping or bucking the system but rather in shedding the framework of optimals that demands either full deference or full retreat.

What does this look like, in practice? It might mean, for example, an active choice of work or calling, in lieu of searching out the *best*— a choice that rises above the claim that choosing should be left to the machines or that it doesn't matter anyway. It might look like a commitment to a belief, a faith in something that defies both calculated utility and nihilistic escape. It might be the choice to dream up new myths, new stories, when the old ones cease to serve. Or maybe it's simply the choice to see the world as a little less subject to our control, a little more worthy of our reverence.

I tend to agree with Newton, even though I'm not particularly religious myself. Our modern-day rationalists have acted in cahoots with the optimizers. They've done us a disservice, by teaching us that only what's in the province of the rational has merit. By proving away religious approaches—alongside religion's various contradictions— they've restricted the frame to that of pure reason.

Or, as the writer Wendell Berry, who coincidentally has lived and farmed for decades just a few miles from Amazon's new Air Hub, puts it, in counter to the rationalists, "The farther we extend the radius of knowledge, the larger becomes the circumference of mystery."

All this said, don't blame the atheists, who are but standing on the shoulders of giants. This all began, ironically and after all, with a very religious impulse, an anxiety about knowing one's fate, and the desire to take that fate, and society's, into our individual hands—to tie daily experience to the self-contained frame of mathematics. The same impulse and anxiety that built up optimization helped undo it.

THAT BLUE SEPTEMBER DAY IN 2020, I DRIVE OFF FROM THE AMAZON Air Hub, down past the horse pastures and strip malls, through the hilly smell of mowed grass, into the Kansas plains and up toward the Rocky Mountains. I spend a night at a campground a few miles off the road. The air at night is getting cold, and I fall asleep to a heavy rain on the camper shell of my truck. The next morning, I stop for breakfast in Missoula and push on, past Spokane, where a brief rainbow emerges to crown the highway. Interstate 90 is a blur, and rain and asphalt mix with the Seattle traffic. Around 5:00 p.m. I emerge from both traffic and gray as the autumn sun bears down on the Skagit Valley. The fields are soft, a spell of sleepiness cast. I stop for gas, I keep on down the road until the very end and, reaching the ferry terminal, settle in for the last leg home.

When I return to the island, after some ten thousand miles driving across the continent and back, there's a quiet as animals burrow and the low-hanging sun prepares to set. I'm far from the sparkle of

San Francisco, far from the bulldozer man carving his ruts into Amazon's field. It's from this vantage point, looking out—like anyone who's lived a stretch on an island, with a combination of relief and longing—toward the mainland on a crisp autumn day, that I can pause to wonder what comes next.

I set off with a feeling of awe for mathematics, with its potential to bound and search the world and find what's best. And then somehow, after many stories and many miles, I ended up on an island, in a manner and fashion impossible to calculate in advance. It'd be a tidy story to say I opted out, I refound place, I settled. But that's only half the truth.

The whole truth is that I'm not Roger or Jason. I'm not Bob or Candide or Wendell Berry. I'm not a hippie or a crusader or a farmer. I didn't grow up hunting or fishing or splitting wood, and only later in life did I set myself the task of learning these things. I'm a wannabe homesteader and an optimizer for too many years.

The further truth is that even here, without bridges and stoplights, Amazon delivers in two days. We're all of us tied in, consuming resources from the mainland. I still take on the occasional consulting project, working with big companies and big ideas; I fly out to far-off places. I still find myself pulled toward these twin poles: the impulse to engineer, to model, to harness, alongside the desire to keep on driving, to retreat beyond the confines of the map. In moments of reflection, I'll even admit that perhaps I romanticized opting out, just as much as I'd once romanticized optimization itself.

Yet it was only once I'd chosen a place to be, and a character, that the story began to make sense.

- - -

THIS BOOK STARTED IN SEARCH OF ALTARS AND RUINS, BESIDE A MAN on a bulldozer. On this blank-slate continent, where we raze our monuments and build anew. We slaughter millions of bison and divvy up land into tidy parcels. A century later we pull down the fences and try to bring the bison back. The curtain falls on the frontier, the railroads set in. The techniques of optimization have given us marvels, factory lines and sky cranes and heated garages. They've given us full calendars, a fraying anxiety, the heart pace of a hamster wheel. And they've left us to wonder: What will we build now?

San Francisco may not be as old as Babylon. Our modern lottery doesn't have us in as tight a chokehold. But as America, belatedly, exits its century, as the world lets go, a bit, of the mythos of optimization, perhaps that mythos can become not something to cling to or deny but a set of productive ruins—a foundation to build, a choice to make.

The newest Air Hub plans have fallen through, I read. "Unfortunately," the PR statement announces, "the Port Authority and Amazon have been unable to reach an agreement." Some blame the government; others, Amazon. They point to existing cracks in the supply chain, to public sector inefficiencies, to private greed. We may be entering a recession, the headlines blare. Don't worry, they say. It's part of the business cycle, the usual setbacks, the ups and downs. We'll keep on building. We'll get things back on track.

But the truth on the ground doesn't lie. Growth has slowed, packages are delayed, there's the feeling of winter ahead.

And maybe, just maybe, there's an altogether different reason for the delays. Perhaps it's not Amazon or the government or a recession that's to blame. Perhaps, instead, it all started slowly, with one small step—an email unanswered, a path through the woods. Maybe somewhere in Kentucky, a bulldozer operator decided, for one short moment, to look away. Maybe he paused on his anthill, gazed at the blue September day, opened his door, and stepped out into it. Maybe he lit a cigarette from the pack in his shirt pocket and started walking. Maybe he walked all the way down the hill, past the strip malls and horse fields, along the highway until he reached the river. Maybe he did this in reality or maybe in his heart. Maybe all of us, in some small way, turned away from the Amazon Air Hubs. Maybe, as I did, we put our cars into gear and drove down one road and then another and to the interstate. Maybe we wandered through the Ozark hills, emerged into the Kansas plains, crossed the Rocky Mountains, and slowly wound back home.

ACKNOWLEDGMENTS

A mountain of thanks are due, not least to Courtney, editor extraordinaire, for her enduring patience and insight. To my agent Gillian, who lit up at the idea of this book and whose enthusiasm and intelligence carried it through. To Manuel for reading drafts in various states of disarray, and for his kindness and depth of intuition. To Jessica, who has cheered this book on since it first took inchoate form a decade ago, and who also brought her enormous heart and mind to early drafts. To the Sloan Foundation, whose generous grant supported research and travel. To my mother and father and grandmother, I am grateful beyond words. To others who read drafts, offered ideas and introductions, and simply listened to me talk: Sash and Alex, Casey, Stephanie, Lane, Laura, Chris in Elko, Rob, Kristi, Daniela, Zoe, and so many more whose names will come to me late at night, long after this list has been set firmly into print.

To the various people I spoke with along the way and who opened their lives and hearts to me, both the names that made it into these pages and those that didn't: thank you.

NOTES

Introduction

2 **"America can't use it all"**: Ching-Ching Ni, "Mangled WTC Steel Bought by China," *Chicago Tribune*, January 26, 2002.

2 **a mind-boggling $100 billion**: See *The Wall Street Journal*'s Markets data about Amazon: www.wsj.com/market-data/quotes/AMZN/financials/annual /income-statement.

2 **within six hundred miles**: Information about CVG's advantages is outlined on the airport website: www.cvgairport.com/about/CVG@Work/air-service -development.

8 **an age of anxiety**: See, for example, Christopher Lasch, *The Culture of Narcissism: American Life in an Age of Diminishing Expectations* (New York: Norton, 1979); Joseph Bottum, *An Anxious Age: The Post-Protestant Ethic and the Spirit of America* (New York: Penguin Random House, 2014) (not to be confused with W. H. Auden's poem "The Age of Anxiety"); William Strauss and Neil Howe, *The Fourth Turning: An American Prophecy—What the Cycles of History Tell Us about America's Next Rendezvous with Destiny* (New York: Crown Books, 1997); and Jane Jacobs, *Dark Age Ahead* (New York: Random House, 2004).

8 **popularity of "pre-apocalypse" TV**: Zara Stone, "Why Millennials Are Obsessed with the Apocalypse," *Forbes*, March 2017.

9 **"North Americans prize efficiency"**: Jacobs, *Dark Age Ahead*, 157.

9 **"the speech of its people":** John Dos Passos, *The 42nd Parallel* (New York: First Mariner Books, [1930] 2000), xiv.

1. Flatland Revisited

13 **best-funded lobbies in Washington, DC:** According to the nonprofit OpenSecrets, which aggregates data on campaign finance and lobbying, in 2022, the five top-spending lobby groups in agricultural production were sugar-related: American Crystal Sugar, the US Beet Sugar Association, American Sugar Alliance, Florida Sugar Cane League, and Southern Minnesota Beet Sugar Co-op. Together, these groups spent about $9 million. By comparison, the next highest crop group, the National Cotton Council, spent $850,000. The entire tobacco industry, including manufacturing and marketing, spent about $22 million on lobbying.

16 **Voltaire's *Candide*:** The novella was originally titled *Candide, ou L'optimisme* (*Candide, or Optimism*). It was quickly translated and released in several languages. Voltaire wrote the novella largely in secret and revised it over the course of several years after its publication.

20 **An Iowa congressman:** In 2008, Iowa Congressman Tom Latham commemorated Borlaug's ninety-fourth birthday by reciting this quote widely attributed to Borlaug's grandfather.

20 **Borlaug's "Green Revolution":** While Borlaug's research focus was on crop breeding, the Green Revolution has come to refer to a broad swath of innovations that led to a rapid increase in agricultural yields. These include the mechanization of planting and harvest, the use of chemical fertilizers, the precision approach to pest and weed management, the bioengineering of seeds, and the control of water via irrigation. Daniel Charles's *Lords of the Harvest: Biotech, Big Money, and the Future of Food* (Cambridge, MA: Basic Books, 2001) describes how both idealism and greed helped propel Borlaug's vision through the late twentieth century and beyond.

20 **"saved more lives":** Gregg Easterbrook, "Forgotten Benefactor of Humanity," *Atlantic Monthly*, January 1997.

22 **get big or get out:** As described in Wendell Berry's *The Unsettling of America: Culture and Agriculture* (Berkeley, CA: Counterpoint, [1977] 2015).

25 **the very food we eat:** Paul Conkin, *A Revolution Down on the Farm: The Transformation of American Agriculture since 1929* (Lexington: University Press of Kentucky, 2008).

28 **"dumping" sugar on the market:** Prior to the North American Free Trade

Agreement's implementation in 1994, sugar imports from Mexico were limited by strict quotas. These were phased out under NAFTA, and in 2008, most quotas and tariffs between the two countries were lifted. In late 2015, the International Trade Commission ruled that Mexico had harmed US producers by selling its sugar at below-market prices in prior years. Following the ruling, minimum prices and new quotas were put in place for sugar imports from Mexico.

33 **healthiest societies in history:** Paul Clayton and Judith Rowbotham, "How the Mid-Victorians Worked, Ate and Died," *International Journal of Environmental Research and Public Health* 6, no. 3 (March 2009): 1235–53.

33 **over fifty billion tons:** E. A. Thaler, J. S. Kwang, B. J. Quirk, C. L. Quarrier, and I. J. Larsen, "Rates of Historical Anthropogenic Soil Erosion in the Midwestern United States," *Earth's Future* 10, no. 3 (March 2022): 10.

34 **heels of increasing fragility:** See, for example, Jared Diamond, *Collapse: How Societies Choose to Fail or Succeed* (New York: Viking Penguin, 2005), and David Graeber and David Wengrow, *The Dawn of Everything: A New History of Humanity* (New York: Farrar, Straus and Giroux, 2021).

34 **when tolerances are too thin:** Conversely, we've seen places where inefficiency helps. Errors or mutations help drive evolution. Downtime stabilizes computer programs and toddlers alike. White space—a hermitage or time apart—leads to some of the most creative inventions. Miles Davis once said, "Music is the space between the notes."

38 **A Vermont farm:** Erik Andrus, "From Seed to Sugar: A Vertically-Integrated Model for Small-Scale Turbinado Sugar Production from Organic GMO-Free Beets—Final Report for Sustainable Agriculture Research and Education Grant FNE11-703," 2012.

40 **a mathematical parable:** Edward Abbott Abbott, *Flatland: A Romance of Many Dimensions* (Mineola, NY: Dover, [1884] 1992).

2. Leaving Las Vegas

49 **headquarters of Zappos:** Information on Zappos's history is drawn from the About Us page on the company's website: zappos.com/about.

50 **through a "branding lens":** See case studies including Frances X. Frei, Robin J. Ely, and Laura Winig, "Zappos.com 2009: Clothing, Customer Service, and Company Culture," *Harvard Business Review*, October 20, 2009; Tony Hsieh, "How I Did It: Zappos's CEO on Going to Extremes for Customers," *Harvard Business Review*, July–August 2010; and Anthony K. Tjan, "Four

Lessons on Culture and Customer Service from Zappos CEO, Tony Hsieh," *Harvard Business Review*, July 14, 2010, https://hbr.org/2010/07/four-lessons -on-culture-and-cu.

52 **in his 1946 address:** John Maynard Keynes, "Newton, the Man," written in 1942, available online at mathshistory.st-andrews.ac.uk/Extras/Keynes _Newton/.

52 **Many biographies exist on Newton:** Biographers of Newton include Gale Christianson, James Gleick, William Newman, and Richard Westfall.

52 **"extremely neurotic":** Richard Westfall, *Never at Rest: A Biography of Isaac Newton* (Cambridge, UK: Cambridge University Press, 1980), 53.

54 **most elusive subatomic particle:** The press release from CERN announces the discovery of the Higgs boson, available at home.cern/news/press-release /cern/cern-experiments-observe-particle-consistent-long-sought-higgs -boson.

55 **focus on one part:** Adam Smith, *The Wealth of Nations* (New York: Bantam, [1776] 2003).

57 **time to complete an automobile:** Charles Sorenson, *My Forty Years with Ford* (Detroit: Wayne State University Press, [1956] 2006).

61 **man in love with games:** Biographical information on Claude Shannon and his ideas can be found in books such as those by James Gleick, *The Information: A History, a Theory, a Flood* (New York: Vintage, 2011); William Poundstone, *Fortune's Formula: The Untold Story of the Scientific Betting System That Beat the Casinos and Wall Street* (New York: Farrar, Straus and Giroux, 2005); and Edward O. Thorp, *A Man for All Markets: From Las Vegas to Wall Street, How I Beat the Dealer and the Market* (New York: Random House, 2017).

62 **a shortish paper:** Claude Shannon, "A Mathematical Theory of Communication," *Bell System Technical Journal* 27, no. 3 (1948): 379–423.

64 **"become more complicated":** Matthew Crawford, *Shop Class as Soulcraft: An Inquiry into the Value of Work* (New York: Penguin, 2009).

65 **Tocqueville's resulting journals:** Alexis de Tocqueville, *Democracy in America* (New York: Penguin, [1835] 2003).

66 **Computer Poker Research Group:** Spencer Murray and Jennifer Pascoe, "Skill Trumps Luck: DeepStack the First Computer Program to Outplay Human Professionals at Heads-Up No-Limit Texas Hold'em Poker," March 2, 2017, www.amii.ca/latest-from-amii/media-release-deepstack/.

67 **"Zappos Theater":** As I finish this chapter, in November 2020, I notice a

friend's online salute to Tony Hsieh, the Zappos founder. I later read the news story, an article by Kirsten Grind in *The Wall Street Journal* (March 26, 2021) titled "Zappos CEO Tony Hsieh Bankrolled His Followers. In Return, They Enabled His Risky Lifestyle." My heart sinks as I read about Tony's tragic death, just a few days ago, after being trapped in an East Coast house fire and a week of slow decay in the hospital. In August, he'd resigned from the CEO role at Zappos. He was forty-six. My friend, who had spoken at some of Tony's Downtown Project events, says, "We'd all lost track of him, and then when he died we realized, none of us really knew him."

69 **history of probability:** Ian Hacking, *The Emergence of Probability: A Philosophical Study of Early Ideas about Probability, Induction and Statistical Inference* (Cambridge, UK: Cambridge University Press, 1975).

69 **"does not appear":** As translated in Leonhard Euler, *The Rational Mechanics of Flexible or Elastic Bodies* (Zurich, Switzerland: Orell Füssli, 1960).

71 **"probabilities wrapped in human flesh":** Karl Taro Greenfeld, "Loveman Plays New Game at Harrah's after Tapping F-16s of Debt," *Bloomberg*, August 5, 2010.

74 **"water flows uphill toward money":** Marc Reiser, *Cadillac Desert* (New York: Penguin, 1993).

77 **humans telling machines:** Peter Drucker, *Post-capitalist Society* (New York: HarperCollins, 1993).

3. High Desert Church

82 **Kondo's popular Netflix show:** *Tidying Up with Marie Kondo*, 2019, on Netflix.

84 **"let all your things":** Benjamin Franklin, *Autobiography of Benjamin Franklin* (New York: Henry Holt and Company, 1916), www.gutenberg.org/files /20203/20203-h/20203-h.htm.

85 **"every man his own priest":** Attributed to Martin Luther.

87 **"pregnant with change":** John Stuart Mill, "The Spirit of the Age," in *Collected Works of John Stuart Mill, vol. xxii, Newspaper Writings, December 1822–July 1831* (London: Routledge, 1981).

87 **describes Mill's writing as:** Thomas Babington Macaulay, *The Miscellaneous Writings and Speeches of Lord Macaulay: Contributions to the Edinburgh Review,* vol. 2 (Project Gutenberg, 2008), www.gutenberg.org/files/2168/2168 -h/2168-h.htm.

88 **"economics are the method"**: Ronald Butt, "Margaret Thatcher: Interview for *Sunday Times*," *Sunday Times*, May 3, 1981, www.margaretthatcher.org /document/104475.

89 **"marketplace of ideas"**: Justice Oliver Wendall Holmes Jr. wrote in his dissent in *Abrams v. United States* (1919) that "the ultimate good desired is better reached by free trade in ideas—that the best test of truth is the power of the thought to get itself accepted in the competition of the market."

91 **Manhattan Project**: Stanislaw Ulam, *Adventures of a Mathematician* (Berkeley: University of California Press, [1976] 1991).

92 **highest-income zip codes**: According to data from census.gov, at about $110,000, the median annual household income in Los Alamos is more than twice that of New Mexico as a whole.

97 **"I was obsessed"**: The biographical background on Marie Kondo has been drawn from Kondo's books as well as sources such as Sarah Kessler, "Decluttering Your Life Is Not Just a Trend—It's Big Business," *Fast Company*, September 2, 2015, and Alison Beard, "Life's Work: An Interview with Marie Kondo," *Harvard Business Review*, May–June 2020.

98 **"more or less alike"**: Alexis de Tocqueville, *Democracy in America* (New York: Penguin Books, [1835] 2003).

99 **"as his juggernaut moved forward"**: John T. Flynn, *God's Gold, the Story of Rockefeller and His Times* (New York: Harcourt Brace and Company, 1932).

99 **Mather claimed that:** See, for example, a description of the Mather and Company posters being sold by the Ross Art Group in New York City: "Mather Work Incentive Posters Celebrate the Spirit of the American Workforce," *Antique Trader* (blog), December 5, 2019, www.antiquetrader.com/collectibles /mather-work-incentive-posters-celebrate-the-spirit-of-the-american -workforce.

101 **worth on the market:** Called the marginalist revolution in 1970s economics. Walras and Cournot were among those to propose the idea of utility-maximizing individuals. From then on, optimization has been part and parcel of economic practice.

102 **"complicated the problems enormously"**: Ulam, *Adventures of a Mathematician*, xxvii.

103 **Barbara Grothus is disassembling:** Clarke Condé, "Shining a Light on a Black Hole: An Interview with Barbara Grothus," *Weekly Alibi*, August 22, 2019, alibi.com/art/59178/Shining-a-Light-on-a-Black-Hole.html.

103 **"They should leave the uranium in the ground"**: Robert Nott, "Los Alamos' Black Hole Is Going Dark," *Santa Fe New Mexican*, July 28, 2018,

www.santafenewmexican.com/news/local_news/los-alamos-black-hole
-is-going-dark/article_a83de44e-0f9d-530c-bd40-1b3b4f6556f5.html.

105 **"I deliberately schedule":** "How to Tidy Your To-Do List Like Marie Kondo,"
Fast Company, April 27, 2020, www.fastcompany.com/90490780/how-to
-tidy-your-to-do-list-the-konmari-way. More recently *People* magazine, citing a webinar, reports that Kondo has "kind of given up" on keeping her
house in perfect order. Alexis Jones, "Marie Kondo Admits She Has 'Kind of
Given Up' on Extreme Tidiness, Says Her House 'Is Messy,'" *People*, January
27, 2023, people.com/home/marie-kondo-admits-she-has-kind-of-given-up
-on-tidying-up/.

4. Metaphoric Breakdown

107 **Households turned up the heat:** The background information on the Texas
crisis has been drawn from articles including "Thousands of 'Cold-Stunned'
Sea Turtles Rescued off Coast of Texas," Reuters, February 18, 2021, www
.reuters.com/article/us-usa-weather-texas-turtles/thousands-of-cold
-stunned-sea-turtles-rescued-off-coast-of-texas-idUSKBN2AI0GZ; Jeremy
Schwartz, Kiah Collier, and Vianna Davila, "Power Companies Get Exactly
What They Want: How Texas Repeatedly Failed to Protect Its Power Grid
against Extreme Weather," *Texas Tribune*, February 22, 2021; and Loren
Steffy, "Texas's Independence Didn't Cause the Power Crisis," *Texas Monthly*,
February 27, 2021.

108 **"First gradually, then suddenly":** Ernest Hemingway, *The Sun Also Rises*
(New York: Charles Scribner's Sons, 1926).

112 **The year 2008:** According to the Edelman Trust Barometer, faith in financial
institutions among Americans age thirty-five to sixty-four dropped from 58
percent in 2008 to 38 percent in 2009. While in other countries, especially
developing countries, banks continued to be associated with a force of pros-
perity, American faith in traditional finance never recovered to its pre-2008
highs.

114 **"It's not a 'tech sell-off'":** Tweet by @Peter_Atwater on May 12, 2022, 5:34
a.m., twitter.com/Peter_Atwater/status/1524699523716722688.

118 **interpret the material world:** George Lakoff and Mark Johnson, *Metaphors
We Live By* (Chicago: University of Chicago Press, 1980).

120 *The Coal Question:* W. Stanley Jevons, *The Coal Question; An Inquiry Con-
cerning the Progress of the Nation, and the Probable Exhaustion of Our Coal
Mines* (London: Macmillan, 1865).

122 **40 percent of the grid:** According to ERCOT, the system was four minutes and thirty-seven seconds away from a total shutdown, which occurs when the system dips below 59.4 Hz for an extended period of time. Had that happened, according to Texas station KXAN's Wes Rapaport, a "black start" situation was likely, meaning a total collapse and weeks to restore power to its twenty-six million customers.

122 **"This crisis situation was unprecedented":** From *The New York Times*, February 16, 2021, article quoting Hogan, and *Harvard Kennedy School Environmental Insights* podcast with Hogan recorded on March 2, 2021.

125 **Exxon wants to change:** Jennifer Hiller, "Oil Majors Rush to Dominate U.S. Shale as Independents Scale Back," Reuters, March 19, 2019, www.reuters .com/article/us-usa-shale-majors-insight/oil-majors-rush-to-dominate -u-s-shale-as-independents-scale-back-idUSKCN1R10C3.

126 **"inexplicable and ungovernable of diseases":** Joan Didion, "John Wayne: A Love Song," in *Slouching Towards Bethlehem* (New York: Farrar, Straus and Giroux, 1968), 25.

5. False Gods

133 **"His great weakness":** Tad Friend, "Sam Altman's Manifest Destiny," *New Yorker*, October 10, 2016.

134 **Stockton Economic Empowerment Demonstration:** Information can be found at www.stocktondemonstration.org.

134 **"economist as plumber":** Esther Duflo, "The Economist as Plumber," *American Economic Review* 107, no. 5 (2017): 1–26.

138 **A breathless *Forbes* article:** Judith Magyar, "Indoor, Vertical Farming Has a Big Role in Sustainable Agriculture," SAP BrandVoice / Forbes, February 2022.

139 **about optimization's reach:** Kurt Vonnegut, *Player Piano* (New York: Random House, [1952] 2006).

140 **As Immelt describes it:** Jeffrey R. Immelt, "How I Remade GE and What I Learned along the Way," *Harvard Business Review* 95, no. 5 (2017): 42.

147 **awareness of climate change:** According to a poll by the AP and Energy Policy Institute at the University of Chicago.

148 **increasing public awareness:** Eugene Linden, *Fire and Flood: A People's History of Climate Change, from 1979 to the Present* (New York: Penguin, 2022).

149 **breakfast cereal produced:** Rainforest Crisp and Rainforest granola were created by Bob Weir and Mickey Hart of the Grateful Dead. The company donated a percentage of profits to rain-forest preservation, and some of the

cereals included rain-forest nuts. Per CEO Steve Bogoff: "In most households, the cereal box is the only package that gets read."

154 **everything as a calculation:** In the wake of the late 2022 implosion of the multibillion-dollar crypto exchange FTX, whose founders identified as effective altruists, the crown philosopher-prince of EA, William MacAskill, sought to distance himself from the fiasco. In a confessional apology delivered on Twitter, MacAskill argued that a strategy of doing the most good possible need not imply that "the ends justify the means." Indeed, he goes on to link to past EA writings on the question of how to maximize maximally—that is, to gain units of "doing good" without losing units by doing bad in the process. Computer scientist and effective altruist Eliezer Yudkowsky sums it up like this: "Go three-quarters of the way from deontology to utilitarianism and then stop. You are now in the right place. Stay there at least until you have become a god" (Tweet from @ESYudkowsky on February 25, 2022).

159 **Shoshana Zuboff coined:** Shoshana Zuboff, *The Age of Surveillance Capitalism* (New York: Hachette, 2019).

6. The Treachery of Optimals

166 **"The sheer number":** "Interview with Montana State University Alum, Jason Baldes, M.S.: Buffalo Advocate," Montana State University, February 2017, https://s3.wp.wsu.edu/uploads/sites/578/2016/12/Feature_Feb2017.pdf.

167 **largest and heaviest land animal:** U.S. Department of the Interior, "15 Facts about Our National Mammal: The American Bison," www.doi.gov/blog/15-facts-about-our-national-mammal-american-bison.

171 **President Grant considered it:** According to J. Weston Phippen, "'Kill Every Buffalo You Can! Every Buffalo Dead Is an Indian Gone,'" *Atlantic Monthly*, May 13, 2016, www.theatlantic.com/national/archive/2016/05/the-buffalo-killers/482349/.

176 **animals and plants:** Skogland describes, on the *Mountain & Prairie* podcast, that when an animal is loaded on a trailer and driven hundreds of miles to a slaughterhouse, it knows what's happening. Not only is the practice inhumane, but the animal's stress translates to meat with poor flavor and nutritional value.

178 **A rancher named Conni French:** Nate Hegyi, "Big Money Is Building a New Kind of National Park in the Great Plains," *NPR Weekend Edition Saturday*, December 8, 2019, www.npr.org/2019/12/08/780911812/big-money-is-building-a-new-kind-of-national-park-in-the-great-plains.

181 **early twentieth-century environmentalists:** Samuel P. Hayes, *Conservation and the Gospel of Efficiency: The Progressive Conservation Movement, 1890–1920* (Pittsburgh: University of Pittsburgh Press, [1952] 1999).

182 **more than one occasion:** In *Natural Rivals: John Muir, Gifford Pinchot, and the Creation of America's Public Lands* (New York: Pegasus, 2019), author John Clayton argues that the rivalry between the two men has often been overstated.

183 **flipped traffic patterns:** Tom Vanderbilt, *Traffic: Why We Drive the Way We Do and What It Says about Us* (New York: Vintage, 2008).

184 **without clear answers:** Fred Pearce, *The New Wild: Why Invasive Species Will Be Nature's Salvation* (Boston: Beacon Press, 2015).

185 **mechanization as an intrusive force:** Leo Marx, *The Machine in the Garden: Technology and the Pastoral Ideal in America* (Oxford, UK: Oxford University Press, [1964] 2000).

188 **"I'd have been lying":** René Magritte, *Selected Writings* (Minneapolis: University of Minnesota Press, 2016).

7. After the Gold Rush

189 **After the Gold Rush:** Borrowed from Neil Young's song "After the Gold Rush" (track no. 2 on *After the Gold Rush*, Reprise Records, 1970), with lyrics offering a haunting vision of salvaging what's left of a postempire world.

198 **idea of island biogeography:** E. O. Wilson and Robert MacArthur, *The Theory of Island Biogeography* (Princeton, NJ: Princeton University Press, 1967).

201 **Industriousness and invention:** James Jerome Hill, *Highways of Progress* (New York: Doubleday, Page, and Company, 1912).

201 **"its striking characteristics":** Frederick Jackson Turner, "The Significance of the Frontier in American History" (a paper read at the meeting of the American Historical Association in Chicago, July 12, 1893, 9), http://nationalhumanitiescenter.org/pds/gilded/empire/text1/turner.pdf.

202 **"He'd of been a great man":** F. Scott Fitzgerald, *The Great Gatsby* (New York: Charles Scribner's Sons, 1925).

202 **productivity at home:** Robert Gordon, *The Rise and Fall of American Growth* (Princeton, NJ: Princeton University Press, 2016).

202 **"The Great Stagnation":** Tyler Cowen, *The Great Stagnation: How America Ate All the Low-Hanging Fruit of Modern History, Got Sick, and Will (Eventually) Feel Better* (New York: Penguin, 2011).

206 **safety we find:** David Graeber, *The Utopia of Rules: On Technology, Stupidity,*

and the Secret Joys of Bureaucracy (New York and London: Melville House, 2015).

207 **"cultures manage their continuation today"**: Jane Jacobs, *Dark Age Ahead* (New York: Random House, 2004), 159.

8. Babylon

216 **subject to the Lottery's hold**: Jorge Luis Borges, "The Lottery in Babylon," in *Ficciones* (New York: Grove Press, 1962).

219 **"get off of this L.A. Freeway"**: Guy Clark, "L.A. Freeway," track 2 on *Old No. 1*, RCA, 1975.

223 **the present theory**: Thomas Kuhn, *The Structure of Scientific Revolutions* (Chicago: University of Chicago Press, [1962] 1996).

224 **the explanatory stool**: The "watchmaker" notion of God ascribed to Descartes leads much more easily, and with less conflict, to a materialist and mechanistic science, whereas Newton's bottom-up empiricism leaves the possibility for, and therefore questions about, God's nuanced intervention. In general, continental philosophers have been more comfortable separating the abstract and self-contained realms of mathematics and religion from the day-to-day, while their British and American counterparts doggedly, and often futilely, have worked to reconcile empirical evidence with intuition and scripture.